The Aquaculture Of Striped Bass

A Proceedings

Joseph P. McCraren
Technical Editor

UM-SG-MAP-84-01

Copy Editor: Mark Jacoby
Cover art: Sandy Harpe

The publication of this technical report is made possible by grant #NA81AA-D-00040, awarded by the National Oceanic and Atmospheric Administration to the University of Maryland Sea Grant Program.

Additional copies of this publication may be obtained by writing:

Maryland Sea Grant Program
1224 H.J. Patterson Hall
University of Maryland
College Park, Maryland 20742

First Printing, 1984
Second Printing, 1985
Third Printing, 1987
Fourth Printing, 1989

Preface

To succinctly develop a list of important publications regarding the culture of striped bass and its hybrids, one would have to begin with the hormone-related work of Stevens in the late 1950s and early 1960s. A void of some 10 years would then follow culminating in the first "bible" on the subject prepared by Bayless (1972), an invaluable contribution at a time when interest in culture of the species was spreading beyond the geographic confines of the southeast. Bishop (1974) then published an important paper on the use of tanks for spawning, and in 1976 Bonn et al. gathered all available information on the subject and produced a manual of cultural guidelines. These publications may be considered the core--foundation, or building blocks--from which culture of the species has evolved, and for that matter, these proceedings.

Based on the positive reaction of attendees to the 1982 finfish conference at Annapolis, coupled with a traditional interest in and need for information on striped bass in the Chesapeake Bay area, it was determined in the summer of 1982 that a second conference be held and subject matter devoted to the striped bass and its emerging hybrids. Beyond the provincial interest and need, it was further recognized that such a conference could serve as a forum, and include information of historical value as well as current, unpublished findings spanning the years since publication of the Bonn et al. manual.

Those of us involved in development of the program are hopeful that these conference proceedings prove to be as valuable to culturists as those publications alluded to earlier. Its contents are unique in that subject matter ranges in an orderly manner from the aforementioned historical aspects of the species to life history per se, culture and production of both the striper and its hybrids in extensive and intensive systems, cryopreservation of sperm, feeds and feeding, diseases and parasites, to unusual but needed information on commercial production, regulatory constraints and marketing.

Successful conferences do not just happen and this one was no exception. They are successful because of the efforts of individuals such as Don Webster and Bill Sieling who convened and guided the program committee during the summer of 1982. The committee consisted of: Bill DuPaul, Merrill Leffler, Howard King, Tony Mazzaccaro, Don Webster, Stewart Tweed, Doug Martin, Andrew Manus, Jim Murray, Ken Allen, Bill Sieling, John Foster, George Flick, Donn Ward, Wendell Ogden and Joseph McCraren. Success also hinges on the quality of speakers and we feel we had the best in their respective fields to contribute to this volume. And certainly, the lasting value of any such venture is tied to its proceedings. For their contributions to the quality of this proceedings a special thanks to Mark Jacoby and Jack Greer.

Joseph P. McCraren
U.S. Fish and Wildlife Service

Introduction

Seldom has the emergence of a new technology been more timely--declining natural stocks and increasing demand have conspired to make the culturing of striped bass highly desirable. And now, as the papers in this proceedings attest, it is not only possible but increasingly practical. In reservoirs, ponds, and streams throughout the country, thriving cultured striper populations--including hybrids--have demonstrated the potential of modern aquaculture technology. Of course larger and more difficult challenges--such as the stocking of an open estuary like the Chesapeake Bay--still remain unanswered, though current state efforts highlight the use of striper hatcheries to bolster dwindling natural stocks.

Such probings into aquaculture would not have seemed quite so justified as little as twenty years ago; 4,000 years of aquaculture history had given the striper effort respectability and hope but too few guidelines. An infusion of modern science was needed to bring the eighty years of striper culture trial-and-error beyond infrequent and fortuitous success.

Striped bass culture debuted in 1879 when fingerlings were seined from New Jersey's Navesink River and transported by train to San Francisco Bay and released. This, and a similar effort in 1881, produced the first sustaining population outside the striper's natural range of the Atlantic and Gulf coasts.

Fast on the heels of this success was the establishment of a prototype hatchery in Weldon, North Carolina, alongside one of the striper's natural spawning grounds in Albemarle Sound. Although efforts to spawn stripers at Weldon succeeded, subsequent attempts in Havre de Grace, Maryland, and on the San Joaquin River in California failed. Interest in striper culture then waned as people realized that success was dependent on the chancy acquisition of ripe brood fish.

The art of striper culture remained virtually without patrons until 1954, when it was noticed that the newly impounded waters of South Carolina's Santee-Cooper Reservoir supported a reproducing population of the normally anadromous striped bass. The stripers were spawning in the reservoir's tributary streams and, for the first time, completing their life-cycle in impounded waters.

The late 1950s became boom years for pond and reservoir construction and the demand for fingerlings to stock them ran high. Because the striper had adapted unaided to Santee-Cooper's fresh waters--and because attempted transplantations of wild adults and fry had failed--interest shifted to pond culture. In 1961, the South Carolina Wildlife Department constructed the Moncks Corner Striped Bass Hatchery on the tailrace of the reservoir to capitalize on the stripers' own efforts at captive reproduction.

The Moncks Corner Hatchery was not immediately successful, however, because the technology was still dependent on the acquisition of ripe brood fish. The following three years would witness the efforts that were to push striper culture into the realm of science: Robert E. Stevens, then a South Carolina state wildlife biologist, artificially induced ovulation by hormone injection. Freed of the dependence on capture of ripe brood fish, striper culture at last became accessible to pond culturists.

Thanks to those early efforts of scientists like Stevens, and subsequent refinements by him and others, striper culture is now a robust--albeit infant--industry. The plusses on its balance sheet are many:

- The striper and its hybrids are stocked in 456 reservoirs and 15 to 20 inland streams in 36 states. The stocking effort has been so successful--and natural recruitment so poor--that more cultured stripers are now caught than wild fish.
- Over 40 million fingerlings are stocked annually.
- Seventeen state and federal and two commercial hatcheries are devoted to the production of striped bass.
- Nineteen hybrids have been produced. The original cross, striped bass female and white bass male (*Morone saxatilis* X *M. chrysops*) is still the most successful: it grows faster and survives better than the striper.
- The striper and its hybrids are cultured extensively and intensively, in ponds, silos, and raceways; their sperm is cryopreserved; their diseases and pests are known; their feeding mechanics are well understood; and their temperature requirements have been profiled.

Though much work remains to be done, striper culture is a going concern, ready in technique, if not in scale, to satisfy an increasing demand for fingerlings. And the demands for what technical editor Joe McCraren calls this "comer" are many: for farm ponds, where families could grow their own stripers; for reservoirs and lakes, where sportfishermen could catch them; and for raceways or impoundments where aquaculturists could raise the fish for sale to restaurants and markets. The biggest demand, of course, continues to present the biggest challenge: the raising of fingerlings for release into the open environment, especially the Chesapeake Bay, where they could reproduce in numbers sufficient to meet the demands of both the recreational and the commercial fishery.

The motivations for advancements in this area are great. Along the Atlantic Coast, for example, the 1974 commercial harvest of 14.7 million pounds declined to less than 5 million by 1981. Several mid-Atlantic states, urged on by the Atlantic States Marine Fisheries Commission, have sharply curtailed or eliminated commercial striper fishing and fifteen states on all three coasts have plans or programs for estuarine enhancement.

Whatever the causes of the dramatic decline in natural striper stocks--and they are poorly understood--the maturation of striper culture ensures that increasing striper production remains possible. The fieldworkers and scientists of striper culture have, in contributing to this proceedings, outlined the structure and potential of this possibility. The timely and expanded application of their techniques defines its only limit.

Contents

Historical Overview of Striped Bass Culture and Management

Robert E. Stevens

Few fish have as much to offer. Consider that the striped bass thrives in fresh, brackish and salt water; in warm, cool and cold water; in oceans, estuarines, rivers, lakes and reservoirs; and eats, with great enthusiasm, a wide variety of fish and shellfish in all those places; and grows at a rate of two to four pounds per year for up to 20 or 30 years.

Wherever it is found in numbers, it becomes the premier sport fish and the preferred table fish. It is caught and enjoyed the length and breadth of the United States and has even found its way to Russia and Japan.

The striped bass is an anadromous fish which must spawn in freshwater streams. Its original range was on the Atlantic Coast from the St. Lawrence River to the St. John's River in northern Florida and in the Gulf of Mexico from western Florida to Louisiana (Raney et al. 1952). In 1879 and 1881, yearling striped bass were seined from the Navesink River, New Jersey and liberated in San Francisco Bay where a reproducing population soon became established.

In 1941 the Santee Cooper Reservoir was created by the construction of a hydroelectric dam on the Cooper River, South Carolina. The dam served to impound 160,000 ares of water and to innundate the portions of the Cooper and Santee Rivers that had historically supported large populations of striped bass. By 1949, a large population of small striped bass was present and by 1952, a sport fishery for the species had developed in the reservoir. It was determined by Scruggs and Fuller (1954) that the population was landlocked, i.e. that the Wateree and

Congaree Rivers above the reservoir were providing the spawning grounds for the species and that conditions within the reservoir were suitable for the growth and survival of the species. During the next five years the population exploded and produced one of the largest populations of striped bass per volume of water ever to be observed (Stevens 1957). The species thrives to this day within the reservoir.

ARTIFICIAL PROPAGATION

Spawning

The first striped bass hatchery was established in 1906 on the Roanoke River at Weldon, North Carolina by the Bureau of Biological Services, the predecessor agency of the U.S. Fish and Wildlife Service. A prototype hatchery was operated by S.G. Worth beginning about 1881. The hatchery is located on the spawning grounds of the Albemarle Sound population of striped bass and is unique in that fish of both sexes can be taken in the large dip nets in the act of spawning. The ripe eggs are fertilized and placed in hatching jars. The larvae hatch around 36 hours later, depending on water temperature, and swim into the aquaria from which they are stocked into rivers and streams when 3 to 5 days old. In recent years, the hatchery has been operated by the North Carolina Wildlife Resource Department.

In 1961, the Moncks Corner Striped Bass Hatchery was built on the tailrace below the Santee Cooper Reservoir by the South Carolina Wildlife Department. The hatchery was built for the purpose of supplying the demand for larvae which had been created by the succesful establishment of striped bass within the reservoir and the excellent sport fishing which it produced. In other words, fish biologists and sport fishermen were interested in establishing such populations in other reservoirs throughout the United States.

The Moncks Corner Hatchery was patterned after the Weldon hatchery but despite a major fishing effort, no ripe females were captured in the tailrace in 1961 and no eggs were taken.

In 1962, hormones were injected into 162 female striped bass in order to induce ovulation. Due to inadequate facilities and ignorance of the phenomenon of overripeness, only 2.64 million fry were produced (Stevens and Fuller 1962).

In 1963, 429 female striped bass were injected with hormones and 13.8 million fry were produced. It was concluded that new holding facilities for the adult fish would have to be constructed to observe and frequently capture female striped bass in order to better define the role of overripe eggs as a source of egg mortality (Stevens et al. 1963).

In 1964 new holding facilities were available, the negative role of overripe eggs was defined, and 100 million larvae were produced from 337 females which had spawned 372 million eggs (Stevens 1964).

The percent hatch of eggs produced in 1962, 1963, and 1964 was 7.3, 17.0 and 31.0 respectively. Although it was suspected as early as 1962 that ovulated ovarian eggs rapidly deteriorate in quality due to hypoxia, the phenomenon was not demonstrated conclusively until 1964. A technique of intra-ovarian sampling of ova from the ovary of living fish with a small glass catheter was perfected. This enabled the prevention of the overripe state because eggs from the female could be periodically examined microscopically in order to predict the time of ovulation.

The protocol for successful hormone-induced ovulation described below launched striped bass culture in the United States (Stevens 1966).

Protocol for Hormone Induced Ovulation of Striped Bass

1. Adult male and female striped bass are captured during the spring spawning season on or near their natural spawning area and immediately injected with at least 127 I.U. of chronic gonadotropin per pound.
2. Twenty-two hours after injection an intra-

ovarian sample of eggs is taken and examined microscopically in order to predict the time when the eggs will reach Stage VI.

3. Females are segregated according to estimated time of ovulation.

4. After reaching Stage VI, the stomach of the female is palpated every 30 to 60 minutes in order to determine whether ovulation has taken place. Eggs remaining in the ovary more than 30-60 minutes become overripe and will not hatch.

5. Ripe females are manually stripped of their eggs. Sperm from one or more males are stripped upon the eggs and the sex products are mixed by hand.

6. Water is then added and the fertilized eggs are placed in hatchery jars at the rate of 100,000 per jar.

7. The eggs hatch in about 36 hours depending upon the water temperature and are shipped to rearing stations between day 1 and day 5.

At the present time, 17 striped bass spawning hatcheries exist in the 12 states. Most are located far inland and they depend upon spawning stock which were originally introduced into inland reservoirs as fingerlings.

In 1974, Dave Bishop of the Tennessee Wildlife Resource Agency developed a new technique for spawning and hatching striped bass which has been adopted by several of the above-mentioned hatcheries (Bishop 1964). The new technique involves the use of circular tanks into which hormone-injected, wild, gravid males and females are placed. In most cases, the fish spawn, whereupon they are removed from the tank. The fertilized eggs hatch in the tank, and 4 or 5 days later the fry become concentrated around the edge of the tank and are dipped out for distribution to facilities for rearing.

This technique has the following advantages:

1. The labor and equipment involved in handling, ova-sampling, spawning and hatching are reduced or eliminated.

2. The hatchery manager does not have to be experienced in predicting the time of ovulation and egg overripeness. In fact, overripeness is avoided.

3. The hatch is equal or superior to the Stevens method.

4. The adult fish can be released in much better condition, perhaps to spawn again.

This technique is highly recommended.

Care and Feeding of Larval Striped Bass:

Striped bass larvae begin feeding 5 days after hatching and on that day they must be offered brine shrimp or zooplankton or be released in earthen ponds where they find naturally occurring zooplankton and insect life. (Experiments concerning the release of fry into the wild demonstrated a low survival and it was concluded in the early 1960s that the rearing of larvae to fingerlings before release into the wild was the most efficient way to produce striped bass populations in reservoirs.)

The larvae are usually transported to the pond site in sealed, plastic bags containing four gallons of water and one-half cubic foot of gaseous oxygen. The bag is placed in a styrofoam container in order to maintain water temperature at 70°F or less because fry younger than 9 days old are intolerant of higher water temperature (Stevens 1967). Between 25,000 and 100,000 larvae can be transported to the rearing site in this manner in time periods up to about 12 hours. Longer trips require less density.

STOCKING OF LARVAE

Rearing

Between 100,000 and 300,000 larvae are stocked per surface acre. The larvae are stocked at 5 to 10 days after hatching. In 1981, average stocking rate nationwide was 150,000 per acre. Earthen ponds, which have previously been dewatered for 2 to 16 weeks, are filled several days prior to stocking the fry.

Organic fertilizer, such as hay and cotton seed meal, soybean meal or chicken manure is placed in the pond prior to and during the 40-50 day growing phase.

The addition of liquid ammonium nitrate and liquid phosphoric acid along with inoculums of Daphnia pulex has help improve production in recent studies. Air lift pumps which circulate the pond water is a recent innovation which can also increase fingerling production.

The great geographical expansion in the rearing of fingerlings makes it difficult to determine the average nationwide production per acre. Wide variation is the rule both within and between hatcheries. However, a satisfactory average production is considered to be about 40,000 fingerlings per surface acre weighing 45 lbs. in a 40 day growing season. Fingerlings are removed from the rearing ponds after 40-50 days because the natural food becomes depleted. If large fingerlings are desired, the same ponds can be refilled and stocked at the rate of 25,000, 2-inch fingerlings per acre. Artificial diets provide the food. Baker (1965), reared 273,000 five-inch fingerlings in an average of 76 days after restocking in 16 ponds at the Edenton National Fish Hatchery. Survival was 82 percent of the Phase I fingerlings stocked.

The fingerlings are stocked in reservoirs and estuaries in order to establish reestablish or enhance striped bass populations.

Intensive Rearing

Larvae--Most attempts to rear larval striped bass intensively have met with rather limited success. However, biologists at the Gulf Coast Research Laboratory in Ocean Springs, Mississippi, have recently perfected techniques to enable the rearing of large numbers of fingerlings intensively. In 1981 biologists of the laboratory reared 623,000 fry to produce 293,000 fingerlings. The fish were reared to 1-1/2 inches in 38 days in a 38,000 gallon intensive culture system and were stocked in Biloxi Bay.

Fingerlings. Fingerlings can easily be trained to take dried diets and have been reared in large quantities in raceways. In Florida, a now defunct commercial enterprise (Marine Protein Corp.) reared large numbers of fingerlings and sub-adults in raceways and silos. The author was the manager in charge of a pond rearing facility in North Florida (Palatka) where striped bass larvae were reared to fingerlings and a raceway-silo facility in south Florida (Homestead) where the fingerlings were reared and sold as advanced fingerlings or reared as food fish. While most of the information engendered is proprietary, the following statements can be made:

1. Over 300,000 fingerlings were reared in 55 gallon drums or 300-500 gallon horizontal race ways.
2. Sixty thousand fingerlings were sold for stocking in ponds and lakes in the U.S., and, on July 2, 1974, three thousand fingerlings were successfully shipped by air to the University of Tokyo.
3. Two hundred forty-four thousand fingerlings were stocked into 6 silos at lengths varying between 2.1 and 7.3 inches. The silos were of different heights and contained between 18,600 and 36,100 gallons of water.
4. Ten thousand pounds of advanced fingerlings (10-11 inches) and over 20,000 pounds of food fish (14-18 in.) were produced in 6 silos in 1973 and 1974.
5. Major problems included pesticide poisoning (heptachlore epoxide) from surrounding truck farms, bacterial diseases (*Flexobacter columeris*, *Aeromonas* sp.), cannibalism and the "oil crunch," which increased pumping costs beyond feasibility.

HYBRIDIZATION

Between March 31 and April 18, 1965, the author, while serving as manager of the Moncks Corner Striped Bass Hatchery, fertilized three different groups of striped bass eggs with sperm from white bass *Morone chrysops*. The cross was successful and the hybrid has proven itself to be a very valuable addition to the overall striped bass program in the U.S. because it almost invariably survives better than striped bass in the larval, fingerling and sub-adult stages. In addition, it grows as well or better than striped bass for the first five years. It does not successfully reproduce itself in the wild and does not live as long as striped bass, and therefore, produces few fish over 20 lbs. Because of its high survivability, a hybrid fishery can be produced and maintained at much less cost and effort than is possible with striped bass (Ware 1974).

Two other species, *M. americana* and *M. interupta*, have also been crossed with striped bass but none has proven to be as vigorous and as successful as the original cross, striped bass female vs. white bass male (Bayless 1972). Recently the reciprocal hybrid of the original cross has gained acceptance particularly in Florida. Bishop (1974) pointed out that the tank culture method is not effective for hybrid fry production because female striped bass will not release their eggs in the presence of white bass males.

CURRENT STATUS

Since striped bass fry first became available on a dependable basis in 1964, some 456 reservoirs have been stocked with striped bass or hybrids. A few reservoirs have received both fish but typically one or the other is stocked. These reservoirs encompass 5.7 million acres of water. This management effort represents 32 percent by number and 57 percent by area of existing reservoirs in this country. The figures are even more impressive when one considers that most of the northern cold water reservoirs have not been stocked with striped bass. A total of 36 states now have inland striped bass fisheries in reservoirs and streams.

There are currently 17 hatcheries producing striped bass and hybrid fry in 12 states. In 1981, 188 million fry were produced of which 99 million were stocked into earthen rearing ponds. Over 40 million fingerlings were produced and subsequently stocked in reservoirs, streams and estuaries.

Natural Reproduction

Natural reproduction of striped bass is known to be occurring in streams tributary to the following reservoirs:

Name	State
Santee Cooper	South Carolina
Kerr	North Carolina
Keystone	Oklahoma
Texoma	Oklahoma-Texas
Whitney-Grandbury	Texas
Mead	Nevada-Arizona
Powell	Utah-Arizona
Cumberland	Kentucky
Millerton	California
McConaughy	Nebraska

Inland streams associated with the above reservoirs as well as those not associated with reservoir populations are as follows:

Arkansas River	Roanoke
Red River	Santee
Colorado	Wateree
Tennessee-Cumberland	Congaree
Brazos	Platte

There seems to be no reason why the Mississippi and Ohio rivers will not eventually support reproducing populations of striped bass.

HARVEST IN RESERVOIRS

Harvest in reservoirs averages about 1lb./acre/year. The range of harvest is from a trace to about 20 lbs./acre/year. Recent exceptional harvest rates have occurred in Cherokee Reservoir, Tennessee, in 1975 when 11.0 lbs./acre were taken and in Smith Mountain Reservoir, Virginia, when 8.7 lbs./acre were caught. In 1958 it is estimated that 20 lbs./acre were taken, from the lower part of the Santee Cooper Reservoir.

CREEL AND SIZE LIMIT

Creel limits range between no limit and 10 striped bass per day and size limits vary between no limits and 20 inches, minimum length. Creel limits for hybrids range between no limit and 25 fish per day.

ESTUARINE ENHANCEMENT

Currently, the states of Florida, Mississippi, New York, Alabama, North Carolina, South Carolina, Georgia, Texas, Louisiana and Maryland have either active estuarine enhancement programs or have engaged in this activity recently. In addition, the states of Virginia, California, Oregon and Washington are planning such programs.

PROBLEMS IN ESTUARIES

Problems recognized in estuaries include overfishing, toxic chemicals, loss of spawning and rearing habitat, impingement and entrainment of young fish on irrigation and power plant screens and insufficient natural reproduction.

PROBLEMS IN INLAND RESERVOIRS AND STREAMS

Problems in inland waters include variable survival of stocked fingerlings, toxic chemicals, overpredation on food supply, predation on trout and salmon and thermal preference. Some biologists are concerned about the potential for natural hybridization.

FUTURE

The future of striped bass enhancement is very bright because of the demand for this great fish. Natural reproduction in reservoirs is limited to the few that have suitable tributary spawning streams. Most reservoirs will continue to be stocked on a put, grow and take basis. Natural reproduction in estuaries is insufficient to supply current demand and stock enhancement will expand tremendously in estuaries within the next two decades.

Even today the demand for striped bass and hybrid fingerlings exceeds the current ability of state and Federal fish hatcheries to supply the fish. New techniques need to be applied to traditional striped bass cultural methods in order to greatly increase production in existing hatcheries. Such techniques as improved pond fertilization and supplemental aeration are now being tested at selected hatcheries.

A second area of needed research is the optimum size and number of fingerlings needed to stock reservoirs or estuaries on a put, grow and take basis. A major cooperative effort is currently in progress in Lake Greeson, Arkansas, to compare the survival of both 2 inch and 6 inch striped bass after stocking. The study will also provide needed information on natural mortality and stocking rates.

Commercial aquaculture of striped bass which failed in the mid-1970s is being tried now in California and on Long Island in New York. In the future, such enterprises may help supply not only striped bass for the table but fry and fingerlings to supplement enhancement stocking by government agencies.

The decade of the 1970s has seen striped bass stock enhancement increase greatly. Striped bass fishing has been expanded from traditional estuarine areas to numerous inland reservoirs and streams. In fact, inland striped bass fishing has surpassed estuarine striped bass fishing and is still growing. One of the more interesting recent developments is the acceptance of the hybrid as an alternative to striped bass as a management tool for both inland and estuarine stock enhancement efforts. This is because the hybrid demonstrates hybrid vigor. It survives and grows better than striped bass and therefore, produces more fish for the money spent. In addition, it does not reproduce itself in the wild and, as such, offers the fish manager good control over the fate of the population he is trying to manage. It does not, however, live long enough to produce the big trophies usually associated with striped bass. Of the 456 reservoirs stocked to date, over 264 have been stocked with hybrids. Hybrids are the fish of choice for the smaller impoundments. The hybrid is also thought to survive better in warm water impoundments and streams.

CONCLUSIONS

What a fish story! What a fish!

REFERENCES

Axon , James R. and David K. Whitehurst, 1982. Striped Bass Management in Freshwater Impoundments with Emphasis on Limiting Factors. Striped Bass Symposium, 112 Annual Meeting, American Fisheries Society. 11p.

Baker, J. E. 1976. The rearing of Hudson River striped bass at the Edenton National Fish Hatchery, 1975. Final report prepared for Consolidated Edison Company of New York, Inc. 16p.

Bayless, Jack D. 1972. Artificial propagation and hybridization of striped bass, _Morone_, _saxatilis_ (Walbaum). Spec. Rept. S.C. Wildl. and Marine Res. Dept. 135p.

Bishop, R. David. 1974. The use of tanks for spawning striped bass (_Morone_, _saxatilis_) Proceedings of the 28th Annual Conference Southeastern Association of the Game and Fish Commissioners.

Bonn, Edward W., William M. Bailey, Jack D. Bayless, Kim E. Erickson and Robert E. Stevens. 1976. Guidelines for striped bass culture. 102p.

Braschler, Eugene W. 1974. Development of pond culture techniques for striped bass. Proceedings of the 28th Annual Conference Southeastern Association of Game and Fish Commissioners.

Massmann, W. H. 1975. Twenty five years of federal-state cooperation for improving sport fishing. Superintendent of Documents, U.S. Government Printing Office, Washington, D.C., 20402. 36p.

McIlwain, T. D. 1976. Annual progress report on striped bass rearing and stocking program--Mississippi. 33p.

Raney, Edward C., Ernest F. Tresselt, Edgar H. Hollis, V. D. Vladykov and D. H. Wallace. 1952.

The striped bass. Bulletin of the Bingham, Oceanographic Collection Peabody Museum of National History Yale University. 177p.

Scruggs, G. D., Jr. and J. C. Fuller, Jr. 1954. Indication of a fresh water population of striped bass in Santee Cooper Reservoir. Proceedings of the Southeastern Association of Game & Fish Commissioners. pp. 64-68.

Stevens, Robert E. 1957. The striped bass of the Santee Cooper Reservoir. Proceedings of the 11th Annual Conference Southeastern Association of Game of Fish Commissioners. 21p.

Stevens, Robert E., and Jefferson C. Fuller, Jr. 1962. A preliminary report on the use of hormones to ovulate striped bass, Roccus saxatilis (Walbaum). Proc. of the 16th Ann. Conf. S.E. Assoc. of Game and Fish Com., Charleston, South Carolina, pp.222-235. S. E. Assoc. of Game and Fish Com., Columbia, South Carolina.

Stevens, Robert E., Otho D. May, Jr., and Herschell J. Logan. 1963. An interim report on the use of hormones to ovulate striped bass, Roccus saxatilis (Walbaum). Proc of the 17th Ann. Conf. S. E. Assoc. of Game and Fish Com., Hot Springs, Arkansas. S. E. Assoc. of Game and Fish Com., Columbia, South Carolina.

Stevens, Robert E. 1964. A final report on the use of hormones to ovulate striped bass, Roccus saxatilis (Walbaum). Proc. of the 18th Ann. Conf. S. E. Assoc. of Game and Fish Com., Clearwater, Florida, pp. 525-538. S. E. Assoc. of Game and Fish Com., Columbia, South Carolina.

Stevens, Robert E. 1965. A report on the operation of the Moncks Corner Striped Bass Hatchery. South Carolina Wildlife Resources Dept. 25p.

Stevens, Robert E. 1966. Hormone-induced spawning of striped bass for reservoir stocking. Progr. Fish. Cultur. 28(1):19-28.

Stevens, Robert E. 1967. Striped Bass Rearing. North Carolina Cooperative Sport Fishing Unit, North Carolina State University, Raleigh. 14p.

Stevens, Robert E. 1974. Current and future considerations concerning striped bass culture and management. Proceedings of the 28th Annual Conference Southeastern Association of Game and Fish Commissioners. 5p.

Ware, Forrest J. 1974. Progress with _Morone_ hybrids in fresh water. Proceedings of the 28th Annual Conference Southeastern Association Game and Fish Commissioners. 11p.

Life History and Biology of the Striped Bass and Striped Bass Hybrids

Mike Freeze

The striped bass (*Morone saxatilis*) is an anadromous fish originally indigenous to the Atlantic Coast of North America and the Gulf of Mexico from Florida to Louisiana (Bonn et al. 1976). Not until 1954, when Scruggs and Fuller (1955) reported that striped bass had spawned in the tributaries of the Santee-Cooper Reservoirs in South Carolina, was it realized that striped bass could complete their life cycle within impounded waters. Since the late 1950s striped bass have been stocked in inland waters throughout the southeastern states in an effort to create a highly desirable sport fishery (Bailey 1975). However, this fishery did not begin to develop until 1965 when Stevens (1966) devised a method for the artificial spawning of striped bass.

Hybridization of fishes in the *Morone* genus also began in 1965 following this breakthrough by Stevens (Ware 1975). Bonn et al. (1976) stated, "The original objectives of the hybridization program were to produce a fish with the desirable characteristics of the parent species, i.e., retain the size, longevity, food habits, fighting ability and food quality from the striped bass, while acquiring adaptability to exotic environments and less stringent spawning habitat requirements from the white bass (*Morone chrysops*)." Subsequently, hybrid striped bass (*Morone saxatilis* x *Morone chrysops*) were produced and stocked in various waters throughout the southeastern states.

HABITAT

"Striped bass are found in a variety of inshore, estuarine, and freshwater habitats depending upon latitude and season" (Setzler et al. 1980). Larval striped bass live in open waters but are dependent upon current for distribution. Juvenile striped bass prefer clean sandy bottoms over mud, gravel or rock bottoms. In estuarine or marine environments striped bass are found along sandy beaches (Bigelow and Schroeder 1953) and rocky shores (Pearson 1931), but usually are associated with some submerged or partially submerged structure (Setzler et al. 1980). In fresh waters, striped bass and hybrid striped bass are found in natural lakes, man-made reservoirs and large rivers. Their distribution is constantly expanding as they are stocked for fisheries management purposes.

MATURITY

Striped bass and hybrid striped bass maturation is dependent upon water temperature (Setzler et al. 1980). The warmer the water the faster sexual maturation occurs. Striped bass males usually mature at age two (Setzler et al. 1980) while striped bass females begin maturing at age three (Lewis 1962). Hybrid striped bass males (original cross; white bass male x striped bass female) reach sexual maturity in one year and hybrid striped bass females begin maturing at age two (Bishop 1968; Williams 1972).

"Egg growth (in striped bass) begins slowly in the summer and fall, but increases rapidly as the spawning season approaches" (Setzler et al. 1980). Eggs for three consecutive seasons may be contained simultaneously in a striped bass ovary but mature ova are distributed uniformly throughout both ovaries (DeArmon 1948). Mature ova are positioned within each ovary so as to permit their release over a short period of time (Lewis and Bonner 1966).

FECUNDITY

Fecundity estimates for striped bass vary considerable by age-group, body weight, and fork length (Lewis and Bonner 1966). Striped bass fecundity estimates ranged from 136,400 to 246,400 eggs per kilogram of body weight (Morgan and Gerlach 1950; Jackson and Tiller 1952; Lewis and Bonner 1966). Williams (1972) reported a fecundity estimate of 352,000 eggs per kilogram of body weight for hybrid striped bass (original cross) that ranged in weight from 0.9 to 3.2 kg.

SPAWNING

Striped bass spawn in fresh or nearly fresh waters from mid-February to late July (Setzler et al. 1980). As many as 50 males may surround a single female during spawning (Worth 1903), which occurs at or near the surface (Woodhull 1947; Calhoun et al. 1950; Surber 1958) and is completed within a few hours (Lewis and Bonner 1966). Eggs are broadcast freely into the water where fertilization occurs immediately.

Spawning peaks may be triggered by a sharp increase in water temperature (Setzler et al. 1980), however, diel spawning patterns are not evident when the striped bass spawning season is considered as a whole. Water velocity and volume are very important in developing a successful striped bass spawn with high and regular flows resulting in the most successful spawns (Setzler et al. 1980). Sudden decreases in water temperature or stormy weather may result in a cessation of spawning (Calhoun et al. 1950).

Hybrid striped bass (original cross) are not sterile (Bayless 1972) and have been observed spawning in the wild (Williams 1972; Ware 1975). However, successful reproduction of any of the *Morone* hybrids has not been confirmed by collection of fry or fingerlings from natural habitats (Bishop 1968; Bonn et al. 1976; Setzler et al. 1980). Natural hybridization of striped bass and white bass, also, has not been documented.

PREADULT PHASE

Embryonic

Fertilized striped bass eggs are spherical, greenish, nonadhesive, and semibuoyant. They hatch from 29 (22°C) to 80 (11°C) hours after fertilization, depending upon the water temperature. Optimum temperatures for survival of striped bass eggs range from 16 to

23°C. Striped bass eggs require approximately 1.5 h at 18°C to water harden (Shannon and Smith 1968).

Fertilized hybrid striped bass eggs (original cross) resemble fertilized striped bass eggs in all respects. However, hybrid striped bass eggs require 3 to 4 h additional time to hatch (Bonn et al. 1976). Optimum survival temperatures and time required to water harden for fertilized hybrid striped bass eggs (original cross) are similar to those for striped bass eggs. Fertilized reciprocal hybrid striped bass eggs (striped bass male x white bass female) are spherical, amber to yellowish-orange in color, adhesive and semibuoyant (Bonn et al. 1976).

Larvae

Newly hatched striped bass or hybrid striped bass larvae derive nourishment from a very large oil globule. Larvae are not able to maintain themselves in the water column until they are three or four days old. Until this time they require sufficient turbulence to keep them from settling to the bottom where they would smother. A velocity of 0.3 m/s is sufficient for fry suspension (Setzler et al. 1980). Striped bass larvae form small schools when approximately 13 mm in length. Striped bass and hybrid striped bass begin feeding at 5 days after hatching at 18°C.

Larval striped bass can acclimate to temperature changes faster than juveniles (Davies 1970); although the temperature range for larval survival, 10°-25°C, is much less than for juveniles, 4.4°-35°C (Setzler et al. 1980). Larval striped bass and hybrid striped bass are routinely tempered for 1 h when stocked in hatchery production ponds (Bonn et al. 1976). Striped bass larvae will achieve better survival and growth in low salinity waters than in fresh water (Bayless 1972; Lal et al. 1977).

"Larval striped bass feed only on mobile planktonic food" (Setzler et al. 1980). Sufficient concentrations of acceptable prey must be available during the critical first several days of striped bass or hybrid striped bass feeding for any larval year class to be successful. Heubach et al. (1963) determined that the occurrence of planktonic species in striped bass stomachs generally agreed with plankton distributions in the environment, which in turn was controlled by salinity. Striped bass and hybrid striped bass shift from a strict planktonic diet to a diet containing small crustaceans and aquatic insects by age 40-50 days (22-35 mm TL).

Juvenile

At age 80-90 days, fish up to 20 mm in length are a preferred food of striped bass and hybrid striped bass 50 to 80 mm TL. Striped bass greater than 150 mm are generally piscivorous (Setzler et al. 1980). Subadult striped bass (1-3 yr) feed primarily in schools at or near the bottom (Bason 1971) with foraging being repeated in the same location until food is unavailable (Bowles 1976).

ADULT PHASE

Habits

Striped bass and hybrid striped bass (original cross) are opportunistic feeders in that "the dominant prey consumed in a particular habitat depends upon availability, which, in turn is regulated by environmental factors" (Setzler et al. 1980). Schooling species are often the dominant prey of striped bass and hybrid striped bass. Numerous studies have shown clupeids (*Dorosoma cepedianum* or *petenense*) to be the principal food item of original hybrid striped bass (Bishop 1968; Williams 1971; Ware 1975) and striped bass (Stevens 1958; Ware 1971, 1975) in freshwater. Although reciprocal hybrid striped bass may also feed heavily on clupeids, they may not be as clupeid dependent as striped bass (Ware 1971).

Striped bass and hybrid striped bass (original cross) may follow and feed on schools of fish (Scofield 1928; Williams 1971) but they are not steady feeders. Normally, most of the striped bass or original hybrid striped bass present will feed on a school of forage fish at about the same time. The amount of feeding varies with the season and the time of day (Bishop 1968; Setzler et al. 1980).

Growth

Striped bass exhibit an increase in size along a north-south gradient since their optimal growth period increases in duration as one progresses to the warmer southern climates (Setzler et al. 1980). Although a similar hypothesis has not been examined for hybrid striped bass, it is quite likely that such a relationship also exists for this fish. Growth rates of striped bass and hybrid striped bass vary considerably from area to area since growth is dependent upon various environmental factors. Several investigators have shown that hybrid striped bass (original cross) grow considerably faster than striped bass during the first two years of life (Bishop 1968; Logan 1968; Williams 1971; Ware 1975). Reciprocal hybrid striped bass have also demonstrated early growth superior to that of striped bass (Ware 1975).

Schooling

Adult striped bass (4 yr +) up to a weight of about 4.5 kg usually congregate in large schools. Larger fish may also school, but striped bass of 13.6-18.1 kg are more often found singly or in small groups (Setzler et al. 1980). Hybrid striped bass (original cross), however, usually remain in schools throughout their lifetime.

Migration

"Striped bass from the Gulf of Mexico and from both extremes of its range along the Atlantic Coast rarely undertake coastal migrations" (Setzler et al. 1980). However, striped bass populations from Cape Hatteras, North Carolina, to New England experience substantial coastal migrations (Vladykov and Wallace 1938; Chapoton and Sykes 1961; Clark 1969). The extent and duration of these migrations are strongly influenced by year class strengths and the time when a striped bass enters such a migration depends upon age and sex (Setzler et al. 1980).

Movements of hybrid striped bass (original cross) in coastal rivers appears to be strongly influenced by seasonal fluctuations in river discharge with fish exhibiting a positive affinity for saline waters (Yeager, in press). Some evidence exists that hybrid striped bass may migrate from one river system to an adjacent river system via a saltwater or brackish water connection (Yeager, in press). However, a substantial coastal migration of hybrid striped bass has not been documented.

Interspecific Competition in New Environments

The introduction of any exotic predator into a new environment could potentially have an adverse effect upon native fish species. The effects of striped bass or hybrid striped bass freshwater introductions upon pelagic forage species present has varied but in most cases it has been insignificant (Bailey 1975). Striped bass or hybrid striped bass introductions have had very few noticeable effects on other native fishes in most reservoirs or rivers to date (Surber 1958; Bishop 1968; Ware 1971; Williams 1975; Bailey 1975; Ware 1975; Setzler et al. 1980; Borkowski and Snyder, in press; Germann, in press; Moss and Lawson, in press; Yeager, in press).

REFERENCES

BAILEY, W. M. 1975. An evaluation of striped bass introductions in the southeastern United States. Proc. Ann. Conf. S. E. Assoc. Game & Fish. Comm. 28:54-68.

BASON, W. H. 1971. Ecology and early life history of striped bass, *Morone saxatilis*, in the Delaware Estuary. Icthyol. Assoc. Bull. 4, 122 p.

BAYLESS, J. D. 1972. Artificial propagation and hybridization of striped bass, *Morone saxatilis* (Walbaum). S. C. Wildl. Mar. Res. Dep., 135 p.

BIGELOW, H. B. and W. C. SCHROEDER. 1953. Striped bass *Roccus saxatilis* (Walbaum) 1792. *In* Fishes of the Gulf of Maine, p. 389-404. U. S. Fish Wildl. Serv., Fish Bull. 53.

BISHOP, R. D. 1968. Evaluation of the striped bass (*Roccus saxatilis*) and white bass (*R. chrysops*) hybrids after two years. Proc. Ann. Conf. S. E. Assoc. Game & Fish Comm. 21:245-253.

BONN, E. W., W. M. BAILEY, J. D. BAYLESS, K. E. ERICKSON and R. E. STEVENS, EDS. 1976. Guidelines for striped bass culture. Striped bass Committee So. Div. Am. Fish. Soc. 103 pp.

BORKOWSKI, W. K. and L. E. SNYDER. In Press. Evaluation of white bass x striped bass hybrids in a hypereutrophic Florida lake. Proc. Ann. Conf. S. E. Assoc. Fish & Wildlife Agencies 36.

BOWLES, R. R. 1976. Effects of water velocity on activity patterns of juvenile striped bass. Proc. Ann. Conf. S. E. Assoc. Game & Fish Comm. 29:142-151.

CALHOUN, A. J., C. A. WOODHULL, and W. C. JOHNSON. 1950. Striped bass reproduction in the Sacramento River system in 1948. Calif. Fish Game. 36:135-145.

CHAPOTON, R. B. and J. E. SYKES. 1961. Atlantic coast migration of large striped bass as evidenced by fisheries and tagging. Trans. Am. Fish. Soc. 90:13-20.

CLARK, J. R. 1968. Seasonal movements of striped bass contingents of Long Island Sound and the New York Bight. Trans. Am. Fish. Soc. 97:320-343.

DAVIES, W. D. 1970. The effect of temperature, ph and total dissolved solids on the survival of immature striped bass, *Morone saxatilis* (Walbaum). PhD Thesis, N. C. State Univ., Raleigh, 100 p.

DeARMON, I. A., JR. 1948. Sampling techniques on the ovary of the striped bass, *Roccus saxatilis* (Walbaum). M. S. Thesis, Virginia Polytechnic Inst., Blacksburg, 40 p.

GERMANN, J. F. In Press. Food habits of *Morone* hybrid bass in Clarks Hill Reservoir, Georgia. Proc. Ann. Conf. S. E. Assoc. Fish & Wildlife Agencies 36.

HEUBACH, W., R. J. TOTH and A. M. McCREADY. 1963. Food of young-of-the-year striped bass (*Roccus saxatilis*) in the Sacramento-San Joaquin River system. Calif. Fish Game. 49:224-239.

JACKSON, H. W. and R. E. TILLER. 1952. Preliminary observations on spawning potential in the striped bass (*Roccus saxatilis*, Walbaum). Maryland Dep. Res. and Educ., Pub. No. 93, 16 p.

LAL, K., R. LASKER and A. KULJIS. 1977. Acclimation and rearing of striped bass larvae in sea water. Calif. Fish Game. 63:210-218.

LEWIS, R. M. 1962. Sexual maturity as determined from ovum diameters in striped bass from North Carolina. Trans. Am. Fish. Soc. 91:279-282.

LEWIS, R. M. and R. R. BONNER, JR. 1966. Fecundity of the striped bass, *Roccus saxatilis* (Walbaum). Trans. Am. Fish. Soc. 95:328-331.

LOGAN, H. J. 1968. Comparison of growth and survival rates of striped bass and striped bass x white bass hybrids under controlled environments. Proc. Ann. Conf. S. E. Assoc. Game & Fish Comm. 21:260-263.

MORGAN, A. R. and A. R. GERLACH. 1950. Striped bass studies in Coos Bay, Oregon in 1949 and 1950. Oregon Fish Comm., Contrib. 14, 31 p.

MOSS, J. L. and C. S. LAWSON. In Press. Evaluation of striped bass and hybrid striped bass stockings in eight Alabama public fishing lakes. Proc. Ann. Conf. S. E. Assoc. Fish & Wildlife Agencies 36.

PEARSON, J. C. 1931. Sport fishing in Chesapeake Bay. U. S. Bur. Fish Fish. Circ. 1, 19 p.

SCOFIELD, E. C. 1928. Striped bass studies. Calif. Fish Game. 14:29-37.

SCRUGGS, G. D., JR. and J. C. FULLER, JR. 1955. Indications of a freshwater population of striped bass, "Roccus saxatilis" (Walbaum) in Santee-Cooper Reservoir. Proc. Ann. Conf. S. E. Assoc. Game & Fish. Comm. 8:64-70.

SETZLER, E. M., W. R. BOYNTON, K. V. WOOD, H. H. ZION, L. LUBBERS, N. K. MOUNTFORD, P. FRERE, L. TUCKER and J. A. MIHURSKY. 1980. Synopsis of biological data on striped bass Morone saxatilis (Walbaum). U. S. Dep. Commer., NOAA, NMFS, FAO Synopsis No. 121, 69 p.

SHANNON, E. H. and W. B. SMITH. 1968. Preliminary observations on the effect of temperature on striped bass eggs and sac fry. Proc. Ann. Conf. S. E. Assoc. Game & Fish. Comm. 21:257-260.

STEVENS, R. E. 1958. The striped bass of the Santee-Cooper Reservoir. Proc. Ann. Conf. S. E. Assoc. Game & Fish. Comm. 11:253-264.

_________. 1966. Hormone-induced spawning of striped bass for reservoir stocking. Prog. Fish-Cult. 28:19-28.

SURBER, E. W. 1958. Results of striped bass (Roccus saxatilis) introduction into freshwater impoundments. Proc. Ann. Conf. S. E. Assoc. Game & Fish. Comm. 11:273-276.

VLADYKOV, V. D. and D. H. WALLACE. 1938. Is the striped bass (Roccus lineatus) of Chesapeake Bay a migratory fish? Trans. Am. Fish. Soc. 67:67-86.

WARE, F. J. 1971. Some early life history of Florida's inland striped bass, Morone saxatilis. Proc. Ann. Conf. S. E. Assoc. Game & Fish Comm. 24:439-447.

_________. 1975. Progress with Morone hybrids in fresh water. Proc. Ann. Conf. S. E. Assoc. Game & Fish. Comm. 28:48-54.

WILLIAMS, H. M. 1971. Preliminary studies of certain aspects of the life history of the hybrids (striped bass x white bass) in two South Carolina reservoirs. Proc. Ann. Conf. S. E. Assoc. Game & Fish. Comm. 24:424-430.

_________. 1972. Preliminary fecundity studies of the hybrid (striped bass x white bass) in two

South Carolina reservoirs. Proc. Ann. Conf. S. E. Assoc. Game & Fish. Comm. 25:536-542.

WOODHULL, C. 1947. Spawning habits of the striped bass (*Roccus saxatilis*) in California waters. Calif. Fish Game. 33:97-102.

WORTH, S. G. 1903. Striped bass hatching in North Carolina. Trans. Am. Fish. Soc. 32:98-102.

YEAGER, D. M. In Press. Ultrasonic telemetry of striped bass x white bass hybrids in the Escambia River, Florida. Proc. Ann. Conf. S. E. Assoc. Fish and Wildlife Agencies 36.

Culture Requirements for Striped Bass

Nick C. Parker

Striped bass, *Morone saxatilis*, have been of interest to U.S. fish culturists since the late 1800s when the fish were moved by train from the East Coast to the West Coast. Interest in coastal stockings to increase natural populations was high during the late 1800s, but then waned until recently. The impetus to re-examine striped bass stocking programs for coastal waters was provided by the development of inland striped bass programs in the southeastern states. Such programs now exist in 28 states and the fish have been stocked in 429 reservoirs (Stevens, personal communication). Knowledge of the conditions required for the culture of striped bass and the hybrids of striped bass x white bass, *M. chrysops*, has increased tremendously since Stevens (1966) developed reliable spawning techniques by using hormone injections.

Even though the first striped bass hatchery was constructed in the 1890s at Weldon, North Carolina, knowledge of cultural requirements of striped bass is based largely on less than 20 years of experience. Striped bass culturists have continued to be surprised by the performance of this anadromous fish. Whether fish are cultured in ponds, tanks, or raceways, their cultural requirements can be examined by considering limiting factors in the environment.

Environmental factors that limit the survival, growth, or reproduction of striped bass may be classified as either physical, chemical, or biological as defined for other warmwater species (Parker and Davis 1981). I here review the physical, chemical and biological requirements of striped bass and evaluate these requirements in terms of current hatchery design and operation.

PHYSICAL REQUIREMENTS

Physical factors that may limit the production and survival of striped bass include temperature, photoperiod, light intensity, water clarity, water depth, water velocity, and substrate. In the culture of the species these factors are usually predetermined by the location and design of the hatchery.

Temperature

It is feasible to either heat or cool water a few degrees to facilitate the hatching of eggs. Water temperature should be about 18-20° C for hatching, but can be increased to 30° C or higher after the larvae are 10 days old. Since most striped bass are now reared in ponds, it is rather impractical to attempt direct thermal control. However, water temperature can be manipulated if the ponds receive thermal effluents from power plants, or geothermal water, or are covered with either opaque or translucent plastic, as are some ponds used by tropical fish producers. To my knowledge, no striped bass ponds have been designed with plastic covers, but for small nursery ponds it appears to be an attractive approach. The use of air-inflated plastic covers should prevent rapid changes in temperature and the resultant loss of fry. Warming the ponds in the early spring should also enhance zooplankton populations required as forage for striped bass fry.

Raceways and tanks normally remain at about the same temperature as the inflowing water. In small intensively stocked units it may be practical to heat or cool the water and reuse it several times before discharging it. Lewis et al. (1981) found it economical to control the temperature with electric heaters in a small water reuse system capable of producing 105,000 striped bass fingerlings. In their system well-water at 14° C was heated to 25° C to promote rapid growth and was then lowered to 18-20° C to facilitate harvest of the fingerlings.

Photoperiod

Photoperiod is another physical environmental factor that could possibly affect production and survival of striped bass. Lights are frequently placed throughout the hatchery and operated for the convenience of hatchery personnel. Striped bass culturists also use lights placed just above the water surface and operated during the night to concentrate zooplankton and aid in estimating zooplankton populations. In hatchery ponds, the larvae and fry, being positively phototactic, are presumably attracted by the lights and concentrated with their food supply. However, such concentrations in small areas could increase the incidence of cannibalism if zooplankton is not abundant. Supplemental feeding may reduce the incidence of cannibalism, but hatchery personnel normally feed fish only during the hours of daylight. Striped bass have been reported to forage throughout the 24-hour diel period; the degree to which an artificial photoperiod may alter the foraging of striped bass in ponds is unknown.

The degree to which photoperiod affects the reproductive cycle of striped bass is also unknown. Judging by evidence in other animals, including mammals, birds, and fish (Palmer 1976) one would expect that photoperiod has a profound influence on gonadal development and the time of ovulation. There is evidence in goldfish, *Carassius auratus*, that the effectiveness of hormone injection used to ovulate fish is related to the time of day when the injection is given (Spieler 1977). Gonadal development was greater in male goldfish injected during the morning than in those injected during the afternoon, whereas the response of females was greater when they were injected during the afternoon. It is reasonable to expect that the time of day and the length of the photoperiod influence the successful ovulation, fertilization, and development of striped bass eggs.

Light Intensity

Light intensity affects the behavior and possibly the survival of fry. Bonn et al. (1976) recommended that direct sunlight be avoided when fry are being packaged for shipment. They also recommended that fry be removed from the shipping containers under subdued lighting. Many hatchery managers now commonly stock the fish in ponds before dawn.

Since fry are positively phototactic, they concentrate around the edges of glass aquaria or other translucent tanks, whereas they remain more uniformly distributed in opaque, dark colored tanks. The concentration of fish along the edges of a tank or into the corners of aquaria increases the likelihood of cannibalism. Survival is usually greater in ponds with a plankton bloom than in ponds with clear water. As water clarity increases, the amount of zooplankton forage available to fry decreases. Some culturists are now using dyes in ponds to reduce water clarity and others are using black plastic shades at the water surface. Rees and Cook (in press) reported that striped bass fry survival was higher in shaded aquaria than in non-shaded aquaria. Survival in aquaria also increased as water clarity was reduced by dyes and turbidity. They also found that survival of striped bass in ponds was significantly correlated with the degree of pond shading, but that water clarity did not have a significant effect on survival. It is not known whether reduced light intensity and water clarity in ponds actually decreases the physiological stress of striped bass or merely reduces the incidence of cannibalism and predation.

Water Velocity

Striped bass larvae and fry are very fragile and are easily damaged if swept against the tank wall by the water current. The water velocity and turbulence in the tank should be controlled to avoid the impingement of fish on the effluent screens and to prevent their being swept into the tank walls. Water velocity should be maintained at the lowest needed to prevent water quality deterioration.

In tanks and raceways the water velocity can be increased sufficiently to aid in cleaning the tanks without harming fingerlings and larger fish. Striped bass are rheotactic and swim against a current. In ponds the fish concentrate at the area of freshwater inflow, and also swim against the current if water in the pond is being circulated.

Substrate

Striped bass do not normally contact the substrate, but stay above it in the water column. However, larvae and fry have been observed swimming directly into the walls of aquaria. Fingerlings and larger fish in raceways and tanks may contact the tank walls and bottom if fish density and water velocity are both high. A smooth tank wall surface reduces damage to fish. The smooth substrate also facilitates the cleaning of tanks and raceways.

CHEMICAL REQUIREMENTS

The chemical factors that may limit the production and survival of striped bass include dissolved oxygen, nitrogen gas supersaturation, carbon dioxide, pH, nitrogenous metabolites, inorganic ions, and contaminants. The balance and mix of these chemical factors constitute the aquaculturist's definition of water quality. Ideally, hatcheries are located in areas where water quality is optimum for fish production. However, not all hatcheries are ideally sited, and even those that are ideally sited often experience periods when chemical factors become stressful to fish.

Dissolved Oxygen

Dissolved oxygen (DO) is the chemical factor that most frequently becomes limiting in striped bass hatcheries. It should be maintained as close to saturation as possible and at a concentration no lower than 4 mg/l (Bonn et al. 1976). Striped bass eggs have hatched when DO concentrations were 1.65

mg/l, but development was retarded when oxygen tensions were low (Westin and Rogers 1978). Striped bass fingerlings in ponds have been reported to survive levels of DO as low as 1.4 mg/l (Parker 1979), but the fingerlings were undoubtedly stressed.

Heavily fertilized ponds often develop intense phytoplankton blooms that supersaturate the water with oxygen as a result of photosynthetic activity. These same blooms can cause an oxygen depletion at night as a result of respiration. Since many hatchery ponds are thermally stratified during the growing season, fish and plankton may concentrate in the upper layer of water. Mixing the upper and lower layers of water to prevent stratification tends to produce a uniform concentration of oxygen throughout the pond.

Airlift pumps have been used at the Southeastern Fish Cultural Laboratory to circulate and destratify pond water. They have also been installed at the Edenton (North Carolina) National Fish Hatchery, the Natchitoches (Louisiana) National Fish Hatchery, and the San Marcos (Texas) Fish Cultural Development Center. Ponds with airlift pumps have typically been uniformly mixed and DO concentrations have been almost identical at the pond surface and bottom. In ponds stocked with 50,000 channel catfish, _Ictalurus punctatus_, per hectare, emergency aeration was required for only 20 hours during a 360-day period, where airlift pumps were continuously operated, as compared with 940 hours in control ponds (Parker 1981).

Gas Supersaturation

Before the widespread adoption of airlift pumps in ponds used to culture striped bass can be recommended, a recently identified problem must be further investigated. Colt and Westers (1982) showed that hatchery aeration systems used in coldwater hatcheries in Michigan sometimes produced nitrogen supersaturation. Cornacchia and Colt (in press) found that total gas pressures as high as only 102.9%

can result in both overinflation of the swimbladder and formation of intestinal bubbles in larval striped bass. They also reported that when fed and unfed fish were exposed to identical gas pressures, intestinal bubbles formed in unfed fish but not in fed fish. It is not known to what extent gas supersaturation occurs in warmwater ponds, either with or without airlift pumps. At the Southeastern Fish Cultural Laboratory we have recently constructed several of Bouck's (1982) gasometers and expect to monitor total gas pressures in ponds, both with and without airlift pumps, in 1983.

Carbon Dioxide and pH

Carbon dioxide (CO_2) and pH are two other chemical factors of concern to striped bass culturists. Carbon dioxide is produced as a by-product of metabolic energy production. When excreted into the water, CO_2 combines with it to form carbonic acid, H_2CO_3. An increase in the concentration of carbonic acid increases the hydrogen ion concentration and thus lowers the pH.

Bonn et al. (1976) recommended that pH remain above 6.8 and below 10. An analysis by N.C. Parker and J.G. Geiger (unpublished data) of 57 water samples collected from striped bass hatcheries showed that the water from the 11 most productive hatcheries was slightly basic (pH, 7.3) and well buffered (mean alkalinity, 195 mg/l), whereas that from the 46 other hatcheries was acidic (pH, 6.4) and not as well buffered (mean alkalinity, 81 mg/l).

When dense plankton blooms develop in ponds with low alkalinity, rather drastic diel shifts in pH may result. During the day the phytoplankton removes CO_2 from the water to support photosynthetic activity and the pond pH may reach 10 or 11. During the night planktonic respiration may produce CO_2 in excess of the buffering capacity of the pond, and pH may decline to less than 6. Diel pH shifts can be reduced by adding lime ($CaCO_3$) to the pond to increase the buffering capacity, or preventing the development of dense phytoplankton blooms.

Nitrogenous Metabolites

Ammonia, the primary nitrogenous metabolite produced by fish, results from the deamination of amino acids and is excreted into the water. The un-ionized form of ammonia (NH_3) is much more toxic than the ionized form (NH_4^+). The balance between un-ionized and ionized forms depends on both temperature and pH. At a pH of 9.4 and at 20° C, half the ammonia is in the NH_3 form and half is in the NH_4^+ form. At pH 7 nearly all (99%) of the ammonia is in the NH_4^+ form and at pH 12 nearly all (99%) is in the NH_3 form. As temperature increases, the NH_3 and NH_4^+ ratio shifts toward an increase in the concentration of NH_3.

Ammonia is converted by bacterial action into nitrite (NO_2), and nitrite is converted into nitrate (NO_3). These conversions are made by the bacteria *Nitrosomonas* and *Nitrobacter*, found in the biological filters of water reuse systems and in the bottom substrate of ponds. Even sublethal levels of ammonia have been shown to reduce growth and to cause tissue damage in fish. Although NH_3 has been recognized as the most toxic form, Colt and Armstrong (1981) presented evidence that the NH_4^+ form may also cause tissue damage and be toxic when pH is low. They also noted that levels of un-ionized ammonia of 50-100 mg/l have been detrimental to most aquatic organisms. It is generally recognized that ammonia values should be kept as low as possible. Bonn et al. (1976) recommended that ammonia concentration not exceed 0.6 mg/l for striped bass culture.

Nitrites and nitrates are not usually a problem in raceways or ponds but may become problems in water reuse systems. Nitrite readily complexes with hemoglobin, producing brown-blood disease or methemoglobinemia, which reduces the ability of hemoglobin to transport oxygen.

Organic and inorganic fertilizers are commonly added to ponds to provide the nutrients necessary to support phytoplankton and zooplankton. The protein in fertilizers decomposes to ammonia, nitrites, and nitrates. If ponds are heavily fertilized these nitrogenous compounds may be present at concentrations high enough to be harmful to striped bass. The diel pH shift alters the relative toxicity of ammonia and nitrites throughout the 24-hour period.

Inorganic Ions

Calcium chloride ($CaCl_2$) and sodium chloride (NaCl) have been shown to reduce the toxicity of nitrite to fish. Wedemeyer and Yasutake (1978) found calcium chloride to be more effective in reducing nitrite toxicity in steelhead trout, *Salmo gairdneri*, whereas Tomasso et al. (1980b) found sodium chloride to be equally effective in channel catfish. These differences may be attributed to the physiological differences in anadromous and non-anadromous species. The protection that selective ions provide against the toxicity of nitrogenous compounds in trout and channel catfish has not been observed in striped bass.

Tomasso et al. (1980a) found that an addition of 10 g NaCl and 25 mg MS-222 (tricaine methanesulfonate) per liter of water in the hauling tank prevented the loss of chloride ions from hybrid striped bass. As judged by the increase of corticosteroids in the plasma, these fish were physiologically stressed by the normal netting and handling associated with transportation of the fish. Fish survival 72 hours after hauling was 100% for those treated with salt and the anesthetic, and was 75% for those treated with only the anesthetic. The use of salt and anesthetics in hauling water was also recommended by Bonn et al. (1976).

Contaminants

Westin and Rogers (1978) reviewed the literature and tabulated the toxicity of 31 substances to larval striped bass and the toxicity of 71 substances to fingerlings. Larvae were far more sensitive to contaminants than were fingerlings; the most critical stage of larval development appeared to be at the time of yolk-sac reabsorption. Fitzmeyer et al. (in press) calculated the 48-hour LC50 of simazine to be 5.9 mg/l in 3-day-old striped bass, whereas it exceeded 100 mg/l in 7-day-olds. Sublethal concentractions of contaminants are also stressful to eggs, larvae, fingerlings, and adults. Detrimental effects produced may include abnormal egg development, eye defects, morphological deformities, reduced feeding and swimming ability, impaired gill function, and reduced growth rate (Rosenthal and Alderdice 1976). Striped bass culturists should consider the risks and benefits before applying even commonly used fishery chemicals. For example, even though diesel fuel has been commonly used to control predacious aquatic insects, it can adversely affect striped bass both directly by its toxicity to the fish and indirectly by its toxicity to crustacean zooplankton (Geiger and Buikema 1982), and consequent reduction of forage available.

BIOLOGICAL REQUIREMENTS

The biological factors that may limit the survival, growth, and production of striped bass include pathogens, predators, inter- and intra-species competition, and forage availability. The degree to which these biological factors affect striped bass depends somewhat upon the design and operation of the culture facility.

As fish density increases, the likelihood of infectious organisms spreading throughout the population likewise increases. The stress of handling and crowding also increases the susceptibility of fish to systemic bacterial infections (Wedemeyer 1970). Davis et al. (1982) reported that striped bass were physiologically stressed when exposed to commonly used anesthetics and salt. The new experimental anesthetic etomidate effectively prevented the increase in plasma corticosteroids that is typically used as an indicator of physiological stress. As more is learned about the biological requirements and responses of striped bass to the environment it should be possible to improve the quality of hatchery reared fish.

Inter- and intra-species interactions can also greatly affect survival and performance of striped bass. Cannibalism can be a serious problem in intensively stocked culture units if there is any appreciable size difference among striped bass fingerlings.

The frequency of feeding and the amount of food offered at any one time may influence both growth and survival. If only a small amount of food is offered at each feeding, a few fish may consume all or most of the food and quickly outgrow the fish deprived of food; the larger individuals then begin to cannibalize the smaller fish. Bonn et al. (1976) recommended that fish be graded at least once every 3 weeks to reduce the size differential.

Striped bass larvae and fry are frequently preyed upon by aquatic insects, by other species of fish, and by larger striped bass. The relative survival of 1- to 2-month-old (phase I) and 5- to 6-month-old (phase II) fingerlings has not been adequately determined after fish have been stocked into either inland or coastal waters.

Forage availability is a major factor determining survival of hatchery reared striped bass. When zooplankton populations have been maintained at a high level during the first 2 weeks of pond culture, striped bass fingerling production during a 5- to 6-week period has been relatively high (Geiger, in press); but when zooplankton populations were not maintained during the first 2 weeks, striped bass survival declined. Maintaining the proper forage base appears to be one of the most critical and difficult problems in the management of striped bass.

SUMMARY

Knowledge of striped bass biology and striped bass hatchery management has increased dramatically during the past 20 years. However, pressures on the striped bass resource and the need to augment natural stocks has likewise increased. Resource managers are now seeking ways to enhance the striped bass fishery in both inland and coastal waters. Hatchery managers are able to produce fish, but not enough to meet the demand. Hatchery production can again be increased as a better understanding of the cultural requirements of striped bass is gained.

ACKNOWLEDGEMENT

I thank B.A. Simco and K.B. Davis for reviewing the manuscript. I especially thank all members of the Southeastern Fish Cultural Laboratory for their various contributions to this work and to the research upon which it is based.

REFERENCES

Bonn, E.W., W. M. Bailey, J.D. Bayless, K.E. Erickson, and R.E. Stevens. 1976. Guidelines for striped bass culture. Southern Division, American Fisheries Society. 103 pp.

Bouck, G.R. 1982. Gasometer: an inexpensive device for continuous monitoring of dissolved gases and supersaturation. Transactions American Fisheries Society 111(4): 505-516.

Colt, J.E., and H. Westers. 1982. Production of gas supersaturation by aeration. Transactions American Fisheries Society 111(3): 342-360.

Colt, J.E., and D.A. Armstrong. 1981. Nitrogen toxicity to crustaceans, fish and molluscs. Pages 34-47 _in_ Proceedings of the Bio-Engineering Symposium for Fish Culture. American Fisheries Society, Fish Culture Section, Publication 1.

Cornacchia, J.W., and J. Colt. (_in_ Press). Effects of dissolved gas supersaturation on larval striped bass (_Morone saxatilis_ Walbaum). Journal of Fish Diseases.

Davis, K.B., N.C. Parker, and M.A. Suttle. 1982. Plasma corticosteroids and chlorides in striped bass exposed to tricaine methanesulfonate, quinaldine, etomidate, and salt. Progressive Fish-Culturist 44(4): 105-207.

Fitzmayer, K.M., J.G. Geiger, and M.J. van den Avyle. (_in_ press). Acute toxicity effects of simazine on _Daphnia pulex_ and larval striped bass. Proceedings Annual Conference Southeastern Association Fish Wildlife Agencies 36 (1982).

Geiger, J.G., and A.L. Buikema, Jr. 1982. Hydrocarbons depress growth and reproduction of the cladoceran, _Daphnia pulex_. Canadian Journal Fisheries Aquatic Sciences 39(6): 830-836.

Lewis, W.M., R. C. Heidinger, and B.L. Tetzlaff. 1981. Tank culture of striped bass: production manual. Fisheries Research Laboratory, Southern Illinois University, Carbondale, Project F-2, 6-R. Illinois Department of Conservation. 115 pp.

Palmer, J.D. 1976. An introduction to biological rhythms. Academic Press, New York. 375 pp.

Parker, N.C. 1979. Striped bass culture in continuously aerated ponds. Proceedings Southeastern Association Fish Wildlife Agencies 33: 353-360.

Parker, N.C. 1981. Performance of four aeration devices in channel catfish ponds. Annual Proceedings Catfish Farmers of America Research Workshop 3: 8-9.

Parker, N.C., and K.B. Davis. 1981. Requirements of warmwater fish. Pages 21-28 _in_ Proceedings of the Bio-Engineering Symposium for Fish Culture. American Fisheries Society, Fish Culture Section, Publication 1.

Rees, R.A., and S.F. Cook. (_in_ press). The effects of sunlight intensity on the survival of striped bass x white bass fry. Proceedings Annual Conference Southeastern Association Fish Wildlife Agencies 36 (1982).

Rosenthal, H., and D.F. Alderdice. 1976. Sublethal effects of environmental stressors, natural and pollutional, on marine fish eggs and larvae. Journal Fisheries Research Board of Canada 33: 2047-2065.

Spieler, R.E. 1977. Diel and seasonal changes in response to stimuli: a plague and a promise for mariculture. Proceedings World Mariculture Society 8: 865-882.

Stevens, R.E. 1966. Hormone-induced spawning of striped bass for reservoir stocking. Progressive Fish-Culturist 28(1): 19-28.

Tomasso, J.R., K.B. Davis, and N.C. Parker. 1980a. Plasma corticosteroid and electrolyte dynamics of hybrid striped bass (white bass x striped bass) during netting and hauling stress. World Mariculture Society 11: 303-310.

Tomasso, J.R., M.I. Wright, B.A. Simco, and K.B. Davis. 1980b. Inhibition of nitrite-induced toxicity in channel catfish by calcium chloride and sodium chloride. Progressive Fish-Culturist 42(3): 144-146.

Wedemeyer, G. 1970. The role of stress in disease resistance of fish. Pages 30-35 *in* S.F. Snieszko, ed. A symposium on diseases of fishes and shellfishes. American Fisheries Society Special Publication 5.

Wedemeyer, G.A., and W.R. Yasutake. 1978. Prevention and treatment of nitrite toxicity in juvenile steelhead trout (*Salmo gairdneri*). Journal Fisheries Research Board of Canada 35: 822-827.

Westin, D.T., and B.A. Rogers. 1978. Synopsis of biological data on the striped bass, *Morone saxatilis* (Walbaum) 1792. Graduate School of Oceanography, University of Rhode Island, Kingston, Rhode Island. Maine Technical Report No. 67. 154 pp.

Review of Striped Bass Brood Stock Acquisition, Spawning Methods and Fry Production

Reginal M. Harrell

Artificial propagation of striped bass (*Morone saxatilis*) began as early as 1880 when S.G. Worth (Worth 1882) successfully spawned a ripe striped bass from Albermarle Sound in North Carolina. In subsequent years a successful hatchery was established at Weldon, North Carolina on the banks of the Roanoke River (Worth 1884a,b). This area, at the natural fall line, appeared to be the major spawning grounds for tremendous numbers of striped bass migrating upstream from Albermarle Sound (Worth 1884a). Worth (1884a,b) and his co-workers depended upon commercial fishermen to supply their hatchery with ripe broodfish. This and subsequent hatcheries operated intermittently using similar techniques for many years (Tatum et al. 1966).

Dependence on ripe broodfish caused several other attempts at establishing striped bass hatcheries to fail and efforts were eventually discontinued (Talbot 1966). Pearson (1938) stated that attempts to artificially propagate striped bass at Havre de Grace, Maryland, failed because of inability to procure both ripe males and females at the same time. In 1910, Coleman and Scofield had similar problems on the San Joaquin River in California (Raney 1952).

In 1962, Dr. Robert E. Stevens successfully used hormones to artificially induce striped bass ovulation (Stevens and Fuller 1965; Stevens et al. 1965; Stevens 1967). This breakthrough paved the way for establishing striped bass populations in places where it was never dreamed possible before. The use of hormones negated dependence on sexually ripe fish to successfully operate a hatchery. Because of the work by Stevens (1966,1967) and others, in particular Bayless (1972), Bishop (1975) and the Striped Bass

Committee of the Southern Division of the American Fisheries Society (Bonn et al. 1976), striped bass culture, although still somewhat of an art, can now be duplicated almost anywhere in the world.

There are several excellent and comprehensive publications on the artificial propagation of striped bass and it is advisable that anyone interested in this aspect of striped bass biology obtain copies of one or all of the following: Stevens (1966,1967), Tatum et al. (1966), Bayless (1972), Bishop (1975), and Bonn et al. (1976).

BROODFISH COLLECTIONS

Obviously, the first requisite for striped bass culture is acquisition of adequate numbers of sexually mature brood stock. The same is true for hybridization with other species of *Morone*.

Recent developments in cryopreservation of *Morone* sperm (Kerby et al., in press a,b) offers promise for mitigation of problems associated with collection and retention of live males at a given point in time. Successful cryopreservation will be extremely helpful late in the spawning season when it is frequently difficult to obtain viable males.

For the present, however, biologists are still dependent primarily on collection of broodfish from the wild. Bayless (1972) and Bonn et al. (1976) provided a synopsis of collection techniques and problems associated with each. For purposes of this paper the following will, for the most part, be a restatement of this published information.

Striped bass roe fish should be collected near their natural spawning site and for the most part no sooner than one month prior to their peak spawning times for each particular location (Bayless 1972). Spawning may occur as early as January in Florida (Barkaloo 1967) and as late as June in the Hudson River (Texas Instruments 1977). Spawning temperatures vary over a range of 10° to 25° C with the peaks between 13.9 and 21° C depending on location (Westin and Rogers 1978; Setzler et al. 1980). Bonn et al. (1976) emphasized the importance of suitable gear selection in the capture of broodfish because an inverse relationship exists between the cumulative effects of stress on broodfish and the number of viable eggs produced.

Hook and Line

This method is probably the least desirable means available. Captured fish frequently succumb to stress and lactic acid buildup in muscle tissue. Gravid females are particularly susceptible to mortality associated with hooking. Bayless (1972) states those fish that do survive exhibit characteristics similar to those fish held too long in captivity, i.e., they progress normally toward ovulation to about the one-hour stage, then maturation ceases. Bonn et al. (1976) emphasizes that those fish that are captured by hook and line should immediately be placed in tanks where adequate supplies of oxygen are available. Survival can be enhanced by placing hooked fish in a 1% salt solution.

Gill Nets

Gill nets also can be devastating to sexually mature striped bass and only should be used when other means are not available. Gill nets are effective in catching broodfish, especially white bass (*M. chrysops*) males used for hybridization. Losses resulting from gill nets can be reduced if the nets are "fished" every 15-30 minutes.

Set or Stationary Gill Nets

These nets (including trammel nets) are effective in both lentic and lotic waters. A good feature of these nets is that they can be set at any depth and they cover a substantial area. Bonn et al. (1976) alludes to a preference for monofilament over multistrand nylon but warns that the former is more damaging to captured fish. Bar meshes of two to four inches are usually most effective.

Drift Gill Nets

These nets are generally confined to lotic situations. Drift nets are set perpendicular to the flow and allowed to float with the current. Problems can result with submerged obstructions in the river.

Traps

Traps of various designs have had varying success depending on location and coordination with the timing of spawning runs. Bayless (1972) discusses problems associated with trap location, size and equipment needed to fish the trap.

Haul Seines

Large haul seines have been used with limited success on the Hudson River (Texas Instruments 1977). Seines usually are not effective because of lack of suitable areas near spawning grounds and the depth of the water.

Pound Nets and Fyke Nets

Both of these nets are very effective in capturing brood striped bass. In ideal locations, large numbers of fish can be caught with minimal stress. However, suitable sites to set such nets are frequently considerable distances from the spawning site and the fish may not be advanced enough for ovulation. The expense of procuring and maintaining the nets is also very high.

Hoop Nets

Bonn et al. (1976) states that some states report success using hoop nets in swift, flowing water between 4 and 8 feet deep.

Bow Nets

For many years Weldon Hatchery depended on commercial fishermen using bow nets to supply broodfish for successful operation. These nets resemble oversized dip nets with a hoop approximately six feet in diameter, a bag of five and one-half feet long, and a twenty foot staff (Bayless 1972). Bow nets are effective because they can be placed at a precise depth where striped bass concentrate and capture has minimal effects on the fish (Bayless 1972). The net is fished from a boat which drifts with the current while the hoop is held vertically near the bottom. Water current pushes the boat slightly faster than the net and effectively keeps the net open. When a fish enters the bag, the worker, sensing its pressure through touch, pulls straight up on the staff and entangles the fish (Bayless 1972).

Electrofishing

By far the most successful and least stressful means of collection is by electrofishing. Fish collected are not subjected to frantic struggling during capture and in many cases the fish is narcotized by the shock for one to three minutes during critical picking up and delivery periods (Bayless 1972).

Bonn et al. (1976) describes the basic electrofishing unit as a "...gas powered generator which usually powers an AC, DC, or AC/DC booster, the booster controls the electrical current and current pattern needed to narcotize the fish with various electrode arrangements. The type of generator, booster, and electrodes needed to catch stripers often depends upon the water quality and physical aspects of the collection area.

Probe type electrofishing units utilizing pulsed direct current and variable voltage are ineffective in deep water but may be effective along river banks and relatively shallow areas. Boom-type electrofishing units utilizing alternating current are more suitable in deep areas and tailwater situations. Long stainless steel cables or similar materials serving as electrodes can be insulated to direct the current to the desired depth, usually near the bottom. The unit is capable of covering a large cross-sectional area at a given moment and a large volume of water in a relatively short period of time" (Bayless 1972).

HANDLING OF BROODFISH

Bayless (1972) states that captured broodfish are usually very docile and can be held in a variety of containers with adequate oxygen and temperature. In any situation, handling of captured broodfish should be minimized. Over-handling can result in failure to ovulate or even death.

PRODUCTION OF LARVAE

Until the breakthrough with hormone usage the number of individual states that could have a hatchery was very limited due to their dependence on naturally ripe broodfish. However, since Stevens' (1966, 1967) hormone work and the contribution of South Carolina, North Carolina and Virginia as major suppliers, striped bass and/or hybrids have been established in almost 30 states, of which eight have naturally reproducing populations (Axon and Whitehurst, in press).

Since Bayless (1972) refined the pioneering research of Stevens' production techniques, very little has changed in methods used other than the innovative technique of tank spawning by Bishop (1975). In almost all cases broodfish are collected by electrofishing very near the spawning hatchery, and immediately transported to holding facilities at the site. Although each state or agency has its own eccentricities and must modify the procedure accordingly, the following example will serve to explain the generalized technique.

Once broodfish are placed in the holding facilities, fish are segregated by sex to prevent tank spawning. Usually these fish are still in a semi-narcotized state from electric shock and this is a good time to handle the fish again for injection. Stevens (1967; Stevens and Fuller 1965; Stevens et al. 1968) tested a variety of hormones and found that chorionic gonadotropin (CG) yielded the most satisfactory results in inducing ovulation. His recommended rate was an intramuscular injection of 275-300 International Units (IU) per kilogram of body weight. Males may be injected, particularly late in the growing season, with 110-165 I.U. per kilogram to increase sperm production. White bass males are also injected for hybridization at the same rate as striped bass males.

Injected fish are then left alone for 23 to 28 hours after injection. Then a sample of eggs is removed by insertion of a glass catheter of 3 mm OD through the urogenital opening into the ovary. The sample is then examined under a microscope to predict the time of ovulation. As the fish nears ovulation the oil globule begins to polarize. (For excellent photographs on various stages of egg development see Stevens (1966), Bayless (1972), or Bonn et al. (1976). The importance of accurately predicting time of ovulation cannot be stressed enough. Once ovulation occurs, the eggs break away from the ovarian wall and are cut off from parental oxygen supply. Prolonged delay in spawning could result in overripeness and ultimate egg death from anoxia. Stevens (1967) and Bayless (1972) felt that a grace period of one hour was available before massive mortality occurred. Ovulation should occur within 30 to 45 hours after injection depending on water temperature and relation to time of natural peak of spawning.

At the predicted time of ovulation the female is checked by manual palpation of the abdomen. If ovulation has or is occurring the eggs will flow freely with exertion of slight pressure on the abdomen. Experience is needed to tell whether a fish is ready or should be left a while longer. If it is evident that the female has ovulated she is then anesthesized by spraying a 0.1% quinaldine on the gills (or sacrificed).

A modified dry method of fertilization is used in spawning the fish. Eggs are manually stripped into a dry plastic spawning pan, and then semen and water are added together. This simultaneous addition of water and semen provides a better distribution of the sperm in the pan. At least two males are used to insure against possible sterility of one of the males. After fertilization, excess water is decanted and eggs are placed in McDonald hatching jars at a rate of 100,000 per jar. It is necessary to maintain a continuous flow of oxygenated water to keep the eggs in suspension. Hatching will occur normally within

40 to 48 hours, depending on water temperature (Bayless 1972).

Tank spawning was developed by Bishop (1975) when it became obvious that manpower requirements necessary to run a jar hatchery were prohibitive in some situations. Bishop (1975) found that a circular fiberglass tank with a diameter of 1.83 m and a depth of 0.76 m is the most desirable size to use. Water is supplied by two or more 1.3 cm I.D. tubes facing the tank's perimeter at a 15° angle to the water surface. Water flow enters under slight pressure at 30-38 liters per minute which creates a circular velocity of 0.1 to 0.15 meters per second at the perimeter. Water level should be maintained 10 cm from the top and is controlled by a center stand pipe surrounded by a fine mesh screen (0.45 m diameter) with a mesh of 19.7 to 23.6 openings per cm. A perforated air line is fixed to the bottom of the screen such that air can be bubbled around the screen to prevent eggs and fry from clogging the openings.

Females are injected after capture and placed in the tanks. Injection rates are similar to those described by Stevens (1967) except that males,usually two per tank, are not injected until approximately 24 hours prior to the predicted time of ovulation. Bishop (1975) emphasized that it is important that the fish not be disturbed during spawning. Spawning usually occurs within 36 to 62 hours, depending on water temperature. All fish are removed from the tank after spawning when the eggs have water-hardened.

There are advantages and disadvantages to both techniques. The biggest advantage tank spawning has over manual spawning is in manpower usage. Manual spawning usually requires at least a three-man crew to be present 24 hours a day for efficient operation. Tank spawning only requires a skeleton crew to insure against equipment failure as the fish spawn themselves. Also, the experience in accurately predicting ovulation time is not necessary in tank spawning. Tank spawning is much easier on broodfish and there is a higher survival rate compared with manual stripping.

Major disadvantages of tank spawning encompass fry

estimation and removal as well as space requirements. Fry removal is accomplished by scooping and siphoning. Estimation of numbers is usually accomplished by taking several volumetric samples with a long glass tube (1.3 cm I.D.) and counting captured fry or eggs. Manual stripping and placement of eggs into volumetrically graduated McDonald jars affords a much better handle on total production. Fungus buildup on dead eggs in tanks can also be a problem since dead eggs are not removed by water velocity due to differences in specific gravity as occurs in jar culture. Bishop (1975) stated that optimum production would be about one million fry per tank per week, depending on the broodfish size. Manual stripping can turn out 20 to 30 million fry in the same space and time required for 6 million produced by tank spawning. This of course would be dependent on suitable numbers of broodfish. For a more detailed format of tank and manual spawning, see Bayless (1972), Bishop (1975) and Bonn et al. (1976).

HYBRIDIZATION

Hybridization of the genus Morone was first conceived in the early 1960s by the three men who are most responsible for the success of striped bass propagation today; Jack Bayless, then with North Carolina Wildlife Commission, David Bishop, Tennessee Wildlife Resources Agency, and Robert Stevens, then with South Carolina Wildlife Resources Department. The first successful Morone hybridization was in 1965 by Stevens and was a cross between striped bass female x white bass male (Bishop 1968). Since then there have been many crosses within the genus Morone, including back crosses and second generations (Smith et al. 1967; Bayless 1968, 1972; Bishop 1968; Kerby 1972; Ware 1975).

Bishop (1975) reported striped bass and white bass will not hybridize with tank spawning techniques, therefore it is necessary to accurately predict time of ovulation of the female and manually spawn the fish to produce Morone hybrids. Prediction of ovulation of white bass is similar to that of striped bass except all the eggs in white bass do not mature simultaneously as in striped bass (Bayless 1972). Also, it is interesting to note that white bass eggs are adhesive and smaller than striped bass eggs.

FRY SHIPMENT

Shipment of fry and eggs is a relatively simple process using styroform tropical fish shipping containers. Larvae are placed in sealed plastic bags containing water and pressurized oxygen. The bag is then placed in the styroform container, which offers protection from rupture and insulation from temperature changes.

Bayless (1972) maintains prolarvae can be held with minimal mortality for 48 hours at concentrations up to 10,000 per liter of water. The bags should be half filled with water and half filled with oxygen. Survival is increased when the temperature of the water is maintained between 15.5 and 18.3^{o} C. with no more than 1° C. temperature variance during transit. Survival of younger prolarvae (1 to 2 da) is better than older prolarvae (4 to 5 da) (Bayless 1972).

Eggs can be transported in a similar manner at concentrations less than 5,000 eggs per liter of water (Bayless 1972). Eggs should not be allowed to hatch in transit and should not be transported for more than 24 hours.

REFERENCES

Axon, J.R. and D.K. Whitehurst. In Press. Striped bass management in freshwater impoundments with emphasis on limiting factors. 112th Annual Meeting, American Fisheries Society. Hilton Head, South Carolina.

Barkaloo, J.M. 1967. Florida striped bass. Florida Game and Freshwater Fish Commission Fishery Bulletin 4: 1-24.

Bayless, J.D. 1968. Striped bass hatching and hybridization experiments. Proceedings of the Annual Conference of the Southeastern Association of Game and Fish Commissioners. 21: 233-244.

Bayless, J.D. 1972. Artificial propagation and hybridization of striped bass, *Morone saxatilis* (Walbaum). South Carolina Wildlife and Marine Resources Department. 135 pp.

Bishop, R.D. 1968. Evaluation of the striped bass (*Roccus saxatilis*) and white bass (*R. chrysops*) hybrids after two years. Proceedings of the Annual Conference of the Southeastern Association of Game and Fish Commissioners. 21: 245-254.

Bishop, R.D. 1975. The use of circular tanks for spawning striped bass (*Morone saxatilis*). Proceedings of the Annual Conference of the Southeastern Association of Game and Fish Commissioners. 28: 35-44.

Bonn, E.W., W.M. Bailey, J.D. Bayless, K.E. Erickson, R.E. Stevens (eds.) 1976. Guidelines for striped bass culture. Striped Bass Committee of the Southern Division of the American Fisheries Society. 103 pp.

Kerby, J.H. 1972. Feasibility of artifical propagation and introduction of hybrids of the *Morone* complex into estuarine environments, with a meristic and morphometric description of the hybrids. Ph.D. Thesis, Univ. of Va., Dept of Marine Sciences. 172. pp.

Kerby, J.H., L.C. Woods, III, and M.T. Huish. In Press a. A review of methods, advances and problems associated with culture of striped bass and its hybrids. World Mariculture Society Special Publication. 64 pp.

Kerby, J.H., J.D. Bayless, and R.M. Harrell. In Press b. Growth, survival, and harvest of striped bass produced with cryopreserved sperm. Presented 112th Annual Meeting, American Fisheries Society. Hilton Head, South Carolina.

Pearson, C. 1938. The life history of the striped bass, or rockfish, Roccus saxatilis (Walbaum). U.S. Bur. of Fish. Bull. 28(49): 825-851.

Raney, E.C. 1952. The life history of the striped bass (Roccus saxatilis, Walbaum). Bull. Bingham Oceanogr. Coll. 14(1): 5-97.

Setzler, E.M., W.R. Boynton, K.V. Wood, H.H. Zion, L. Lubbers, N.K. Mountford, P. Frere, L. Tucker, and J.A. Mihursky. 1980. Synopsis of biological data on striped bass, Morone saxatilis (Walbaum). NOAA Technical Report, NMFS Circular 433. 69 pp.

Smith, W.B., W.R. Bonner, and B.L. Tatum. 1967. Premature egg procurement from striped bass. Proceedings to the Annual Conference of the Southeastern Association of Game and Fish Commissioners 20: 324-330.

Stevens, R.E. 1966. Hormone-induced spawning of striped bass for reservoir stocking. Progressive Fish Culturist 28: 19-28.

Stevens, R.E. 1967. A final report on the use of hormones to ovulate striped bass, Roccus saxatilis (Walbaum). Proceedings of the Annual Conference of the Southeastern Association of Game and Fish Commissioners 18: 525-538.

Stevens, R.E., and J.C. Fuller, Jr. 1965. A preliminary report on the use of hormones to ovulate striped bass, Roccus saxatilis (Walbaum). Proceedings of the Annual Conference of the Southeastern Association of Game and Fish Commissioners. 16: 222-235.

Stevens, R.E., O.D. May, Jr., and H.J. Logan. 1965. An interim report on the use of hormones to ovulate striped bass (Roccus saxatilis). Proceedings of the Annual Conference of the Southeastern Association of Game and Fish Commissioners 17: 226-237.

Talbot, G.B. 1966. Estuarine environmental requirements and limiting factors for striped bass. Transactions of the American Fisheries Society Special Publication No. 3: 37-49.

Tatum, B.L., J.D. Bayless, E.G. McCoy, and W.B. Smith. 1966. Preliminary experiments in the artificial propagation of striped bass, Roccus saxatilis. Proceedings of the Annual Conference of the Southeastern Association of Game and Fish Commissioners 19: 374-389.

Texas Instruments Incorporated. 1977. Feasibility of culturing and stocking Hudson River striped bass. An overview, 1973-1975. Texas Instruments Incorporated Ecological Services, Dallas, Texas.

Ware, F.J. 1975. Progress with Morone hybrids in freshwater. Proceedings of the Annual Conference of the Southeastern Association of Game and Fish Commissioners 28: 48-54.

Westin, D.T., and B.A. Rogers. 1978. Synopsis of biological data on the striped bass, Morone saxatilis (Walbaum). Univ. Rhode Island Marine Technical Report 67: 154 pp.

Worth, S.G. 1882. The artifical propagation of the striped bass (Roccus lineatus) on Albermarle Sound. Bulletin of the United States Fish Commission for 1981, 1: 174-177.

Worth, S.G. 1884 a. The propagation of the striped bass. Transactions of the American Fish Cultural Association. : 209-212.

Worth, S.G. 1884 b. Report on the propagation of striped bass at Weldon, N.C., in the spring of 1884. Bulletin of the United States Fish Commission 4: 225-230.

Striped Bass Culture at Marion Fish Hatchery

C.J. Turner

Striped bass (*Morone saxatilis*) have been the subject of intensive study since the 1950s, when Scruggs and Fuller (1954) reported a self-sustaining population in a freshwater environment. Propagation of the species became practical after Stevens (1966) developed methods for producing larvae in large quantities. Early efforts to produce fingerlings were not particularly successful, as production was generally low and extremely variable (Sandoz and Johnson 1965; Harper and Jarman 1972; Humphries and Cumming 1973; Rees 1978). Fingerling production increased steadily as techniques became more refined. During the 1970s fingerling production became a viable proposition and numerous state and federal agencies established striped bass propagation programs. In 1976 a hatchery manual, "Guidelines for Striped Bass Culture," (Bonn et al.) was published.

This report is adapted in part from "Guidelines for Striped Bass Culture." Portions of the original text are now considered obsolete, and have been altered to reflect current practice at Marion Fish Hatchery, Marion, Alabama. Culturists should remember that these recommendations are tailored to complement culture conditions at Marion and may not give optimum results elsewhere. Variations of these methods or altogether different methods may give better results at other locations. Culturists are advised to examine all alternatives so that they may identify those which work best at their own hatcheries.

The following discussion is addressed to the propagation of striped bass fingerlings. Hybrid striped bass (*Morone saxatilis* X *M. chrysops*) are very similar to striped bass and recommendations concerning one may also be applied to the other. Many culturists feel that hybrids are less susceptible to environmental stress than striped bass and are therefore easier to culture (Logan 1967).

PRESTOCKING PREPARATION

At Marion Fish Hatchery the objectives of prestocking pond management, in order of priority, are:

(1) to establish a parent stock of desirable zooplankters (cladocerans and copepods) and to promote the expansion of those stocks

(2) to suppress nuisance vegetation while promoting phytoplankton production

(3) to reduce or eliminate predaceous insects and other nuisance animals

(4) to maintain adequate water quality

Zooplankton Management

At Marion striped bass production correlates closely with the production of desirable zooplankters. Ponds which produce more cladocerans and copepods, particularly in the first week after stocking, usually produce more and larger fish (Turner 1981a, 1982). For this reason, and because of implications in stocking mortality, zooplankton management is given top priority at Marion. Other aspects of pond culture, such as nuisance vegetation, disease, and water quality, are also important but are considered secondary to food production.

Zooplankton populations in striped bass culture ponds exhibit characteristic patterns of dominance and succession (Geiger 1981; Turner 1981b). In ponds where fish are abundant, the most desirable zooplankters (particularly cladocerans) peak in abundance shortly after stocking and continually decline thereafter. When fish are scarce, though, cladocerans and copepods usually remain abundant. This indicates that zooplankton population structure is largely shaped by fish predation. The implications of this are clear: the culturist must build up zooplankton populations before the fish are stocked. After stocking, the fish will consume the most desirable plankters as fast as they appear, thus they are unable to increase in number.

At Marion, the weight of striped bass produced is correlated with the length of time which desirable zooplankters persist in the pond. In other words, the longer the food lasts, the larger the fish grow (Turner 1981a, 1982). For those zooplankters to persist under conditions of extreme predation, maximum reproductive and growth rates must be maintained for as long as possible.

The two preceding paragraphs define the objectives of zooplankton management. In order to produce large numbers of healthy, robust fingerlings, the culturist must produce large quantities of desirable zooplankton prior to stocking. The culturist must then foster growth and reproduction of those stocks for as long as possible. At Marion, this is accomplished through selective zooplankton inoculation and judicious use of fertilizers.

The first step in zooplankton management is to insure that a parent stock of high quality zooplankton is present when the ponds are filled. Where water is supplied from infertile reservoirs or deep wells, it may be necessary to add zooplankters from an outside source. Marion ponds are supplied with well water, therefore all receive a zooplankton inoculum as they are being filled. There exists few data on the consequences of inoculation, but preliminary results indicate that it may be of great benefit where initial zooplankton stocks are low. Preliminary results also indicate that quality of the inoculum is as important as quantity. The composition of the inoculum should resemble the composition desired in the pond. Medium-sized, soft-bodied cladocerans (example: *Ceriodaphnia*, *Scapholeberis*) and gravid copepods should be abundant in the inoculum; small-bodied or spiney rotifers (example: *Brachionous quadridentata*, *Polyarthra*) should be less abundant. Unicellular, green algae should be abundant but the inoculum should never contain filamentous algae.

At Marion, special ponds are set up solely for

the purpose of developing inoculum. These ponds are located at strategic points on the hatchery so that each striped bass pond has direct access (via pump or siphon) to an inoculation pond. The inoculation ponds are set up in February or early March. Each is filled, fertilized with organic and inorganic fertilizers, and inoculated with an initial population of cladocerans and copepods. At Marion, each inoculation pond is also stocked with 25/ha adult male bluegills (*Lepomis macrochirus*) to control fairy shrimp (*Streptocephalus sealii*). Grass carp (*Ctenopharyngodon idella*) may also be stocked when circumstances dictate.

Cladocerans and copepods are more efficient foragers than are other common zooplankters (i.e., rotifers). If allowed sufficient time, cladocerans and copepods usually outcompete the rotifers and become dominant (Allan 1976). At Marion, 6 to 8 weeks seems to be adequate for this purpose. In mid-April, when the inoculum is first needed, the inoculation ponds usually support substantial populations of these crustaceans. The inoculum is then pumped or siphoned into the culture ponds as they are being filled. When the inoculation pond is nearly empty, the bluegills are removed and it is refilled, fertilized, and, after 4 to 6 days, stocked with striped bass larvae.

Since the primary objective is the production of abundant zooplankton prior to stocking, and since most zooplankters grow and reproduce in proportion to their food supply (Edmondson et al. 1962; Wright 1965; McCauley and Kalff 1981), it logically follows that the next step (after inoculation) is to establish a large food base for the zooplankton. This is accomplished by promoting a dense bloom of phytoplankton, bacteria, and small protozoans. The bloom should be established as soon as possible after the ponds are filled and maintained for as long as possible thereafter. This point cannot be overemphasized; it is extremely important to develop and maintain a dense bloom during the early portion of the culture period. In this regard, fertilization strategies are extremely important. At Marion, both organic and inorganic fertilizers are selected for their ability to produce the most phytoplankton and bacteria in the shortest period of time.

Low-protein fertilizers, such as hay or alfalfa cubes, tend to be long lasting and offer good long-term fertility, but they break down slowly and do not enter the food chain quickly (Bonn et al. 1976). High-protein fertilizers, such as cottonseed meal, chicken litter, or meat scraps, tend to break down faster and enter the food chain sooner than the low-protein fertilizers. For this reason, cottonseed meal (41% protein) is used at Marion. In keeping with the strategy of developing a bloom as rapidly as possible, application rates are high and weighted towards the early portion of the culture period. The ponds receive an initial application of 250 kg/ha as they are filled, 100 to 200 kg/ha/week for the next 2 or 3 weeks (as water quality permits), and 100 kg or less/ha/week thereafter. The fertilizer is spread in a thin layer across the bottom or crumbled into the pond to maximize its exposed surface area.

When "Guidelines for Striped Bass Culture" was published in 1976, inorganic fertilizers were not highly recommended (Bonn et al. 1976). Recent development of new fertilizers and fertilization strategies appear to be changing this outlook. Liquid inorganic fertilizers (Metzger and Boyd 1980) hold great promise in striped bass culture.

In central Alabama, as in most other freshwaters, phytoplankton production is limited by phosphorus. The ability to supply this nutrient determines a fertilizer's effectiveness (Boyd 1979). The solubility of phosphorus in granular fertilizers is relatively low; therefore, little of it is immediately available to the phytoplankton (Boyd 1981a). However, the phosphorus in liquid fertilizer is rapidly available to the phytoplankton because it dissolves almost immediately after application. That characteristic implies that liquid fertilizers will produce phytoplankton bloom more rapidly than granular fertilizers, and, at Marion, that seems to be the case (although not yet documented). Liquid ammonium polyphosphate (10-34-0) has been incorporated into fertilization strategy at Marion primarily due to the speed with which it produces a dense plankton bloom.

As with the organic fertilizers, liquid fertilizer application rates are high and the schedule weighted towards the early portion of the culture period. Ponds receive 8 L/ha as they are being filled, 8 L/ha/week for the next two weeks, and 4 L/ha/week thereafter. Liquid fertilizer is much heavier than water and should not be poured directly into the pond. In that case, it will sink to the pond bottom before it can dissolve and much of the phosphorus will be absorbed by bottom muds. Instead, the fertilizer should be premixed with water (10:1, water: fertilizer) before application. Also, the fertilizer stock solution should be thoroughly mixed before each aliquot is withdrawn because the nutrients will settle to the bottom of the tank if the stock solution is left undisturbed. For further information on the use and misuse of liquid fertilizers, readers are referred to Alabama Fisheries Section Leaflet 15 (undated), Boyd (1981a, 1981b), Davidson and Boyd (1981), and Metzger and Boyd (1980).

Fertilizers and application schedules listed above are provisional and do not constitute final recommendations. Future research into striped bass culture will almost certainly alter those schedules. Culturists are advised to keep abreast of current developments in this field rather than becoming locked into any predetermined fertilization regime.

The fertilizer schedules listed above may occasionally produce poor water quality in general and oxygen deficiencies in particular. To prevent fish kills, an intensive water quality maintenance program has been implemented at Marion. Such programs are usually labor intensive and, at hatcheries where manpower or equipment preclude such a program, fertilizer schedules (particularly the organics) should be scaled down to safer levels.

Nuisance Vegetation Control

The adage "an ounce of prevention is worth a pound of cure" is never so appropriate as when applied to the control of nuisance vegetation in striped bass culture ponds. "Guidelines for Striped Bass Culture" lists means by which prevention can be accomplished. No method is 100% successful, and all have disadvantages, but some form of nuisance vegetation control must be employed when inorganic fertilizers are involved.

At this point we arrive at a near contradiction in pond preparation strategy. Zooplankton production, as previously discussed, is assigned the highest priority in pond preparation. Zooplankton production is closely linked to phytoplankton production, therefore phytoplankton production is also assigned a very high priority and primary productivity is actively promoted. However, vegetation control also receives a high priority, which almost inevitably demands that the culturist suppress primary production. This places the culturist squarely in the middle of a dilemma -- he has been asked to both promote and suppress primary productivity at the same time. This by necessity calls for a compromise. The nature of this compromise may be better understood by redefining the second priority in striped bass culture at Marion: the culturist must prevent the establishment of nuisance vegetation during the prestocking period with some management technique which minimizes damage to the phytoplankton.

At Marion, the main line of defense against nuisance vegetation is the phytoplankton bloom itself. Once established, the bloom will shade the pond bottom and prevent nuisance vegetation indefinitely. This strategy has one significant drawback; a dense bloom must be quickly established after the ponds are flooded. If not, unwanted vegetation soon appears on the pond bottom. At that point a race for available nutrients and living space soon develops between the desirable phytoplankton and the nuisance vegetation. If the nuisance vegetation wins the race, it will dominate the pond to the detriment of fish production. If the initial development of the nuisance vegetation is retarded, however, the phytoplankton will ultimately shade the vegetation out of the pond.

Preflooding chemical treatments are frequently used to retard initial weed growth. Snow (1977) recommended a preflooding treatment of Simazine at approximately 11 kg/ha for general weed control. This rate usually gives complete weed control, but also suppresses phytoplankton production and impairs water quality (Boyd 1979). At the other extreme, no preflooding treatment greatly aids primary productivity, but nuisance weeds frequently become so dense as to render a pond worthless. At Marion, the best compromise seems to be Simazine at a reduced rate of 5 kg/ha[(1)]. The Simazine is sprayed on the pond bottom 2 to 3 days prior to flooding. The delay allows the chemical to dry on the pond bottom and form a relatively hard, insoluble crust. This helps keep the Simazine on the bottom, where it is needed, and reduces its concentration in the water column, where it is harmful to the plankton. Because the rate of Simazine is reduced, the treatment is concentrated around the edges and in the shallow end of the pond where it is most useful. Deeper areas of the pond are left untreated.

The preflooding strategy described above will not prevent aquatic weed growth. It is designed to merely retard weed growth during the first week after flooding so that a phytoplankton bloom can be established.

Some culturists may reduce growth of nuisance vegetation by eliminating inorganic fertilizers. Admittedly this limits weed growth, but at the price of reduced phytoplankton production. Due to the importance of zooplankton and phytoplankton production in striped bass culture, this method of weed control is not practiced at Marion.

In spite of the culturist's best efforts, aquatic vegetation sometimes becomes established in one or more ponds. Under these conditions, the culturist should attempt at least partial control, because moderate amounts of vegetation are not always dis-

[(1)] At this time the optimum rate at Marion has not been established - 5 kg/ha is recommended until more specific information is available.

astrous, particularly during the latter portion of the culture period. "Guidelines for Striped Bass Culture" discusses various methods and the reader is referred to that text for details. All of the listed methods, with one exception, are fairly effective. That exception is the stated method of filling a pond immediately before stocking and harvesting 4 to 6 weeks later. The underlying assumption is that a profuse growth of aquatic weeds cannot develop in such a short time. Unfortunately, experience at Marion has shown that 4 to 6 weeks is more than enough time for profuse growths to develop when inorganic fertilizers are liberally applied. Also, where ponds are supplied by wells or infertile reservoirs, this technique does not allow enough time to develop adequate zooplankton populations. For those two reasons, "immediate stocking" is not practiced at Marion to control weeds.

Other Prestocking Considerations

At Marion insect control is achieved with diesel fuel. The fuel is applied at a rate of 15-20 L/ha from 3 to 4 days prior to stocking. This treatment is most effective on calm days, when the oil film is not disrupted by wave action.

Many common insecticides, such as Baytex and Dylox, have also been recommended for insect control (Bonn et al. 1976). Research at Marion (Turner 1981) and elsewhere (Von Windeguth and Patterson 1966; Hurlbert et al. 1972; McCraren and Phillips 1977) has shown that desirable zooplankters (particularly cladocerans) are vulnerable to low levels of such chemicals. This can disrupt the food chain and reduce fish production. Use of insecticides during the prestocking period (and for at least 1 or 2 weeks thereafter) is seldom advisable for this reason, therefore insecticides are not recommended for prestocking insect control.

Water quality during the prestocking period is not as critical as after stocking, but water quality parameters should be within acceptable ranges when the fish are stocked. These ranges, as listed by

Bonn et al. (1976), are: dissolved oxygen above 3.0 mg/L; pH from 6.5 to 9.5; salinity from 0.0 to 10 ppt; hardness, no limits; temperature above 19°C (although there is no recognized upper limit, temperatures above 26°-28°C should be avoided if possible). All ponds should be checked 1 or 2 days prior to stocking to insure that water quality parameters are within established ranges.

STOCKING

Stocking Mortality

The stocking process is perhaps the most important single operation in striped bass culture. Improper or careless stocking technique can result in high initial mortality--75% to 100% of the larvae may be lost within 1 or 2 days of stocking. At Marion Fish Hatchery, excessive stocking mortality was implicated in striped bass crop failure during the 1970s. During the past few years improved stocking techniques have greatly alleviated this problem. The improvement is attributed to a three-point program designed to:

(1) minimize physiological stress
(2) provide an abundant supply of high quality food
(3) prevent overexposure to sunlight

Striped bass of all ages and sizes are vulnerable to physiological stress (Humphries and Cummings 1973; Braschler 1974; Bonn et al. 1976). Stress may manifest itself in an acute or chronic form, either of which can be fatal. Therefore, the goal of the culturist is to reduce stress, which in turn should reduce stocking mortality. It is particularly important to maintain the fish in top physical condition prior to stocking. Dissolved oxygen, temperature, pH, etc., should be maintained within prescribed ranges (Bonn et al., 1976). Holding tanks should be cleaned daily and the larvae must be fed ad libidum at least four times daily and, if possible, six to eight times daily. These recommendations, plus those offered in "Guidelines" and elsewhere in this symposium, should yield healthy, robust larvae which can be safely handled.[1]

All possible precautions should be taken when the larvae are being stocked. Handling should be kept to an absolute minimum. Oxygen levels should be maintained while the larvae are being transferred and tempered. At Marion, an air pump or an oxygen cylinder is used to boost oxygen. The larvae should never be overcrowded; transfer and stocking should be accomplished as rapidly as possible (allowing adequate time for tempering). Also, all hatchery personnel should understand the fragile nature of the species and should be prepared to handle the larvae as delicately as possible.

Mortality rates of many marine fish larvae are strongly, and inversely, correlated with food density (Hunter 1981). Eldridge et al. (1981) found that instantaneous daily mortality of striped bass larvae was in inverse proportion to *Artemia* density. The number of striped bass and hybrid striped bass produced at Marion is dependent upon the quantity and quality of food available during the week following stocking. Ponds in which copepods and cladocerans are abundant during this week typically outproduce those in which crustaceans are scarce (Turner 1981a, 1982). These data do not clearly define the nutritional requirements of striped bass larvae, but they strongly suggest that stocking mortality is associated with inadequate food supplies.

The relationship between food availability and stocking mortality again raises the topic of zooplankton production during the prestocking period. As previously discussed, desirable zooplankters must become abundant prior to stocking so that they may withstand fish predation later in the culture period. In addition, those zooplankters must be abundant at the time of stocking so as to reduce stocking mortality. Thus there are two separate, and equally important,

(1) The larvae, of course, should never actually be handled, touched, netted, or otherwise removed from the water, but instead siphoned, dipped up in cups, etc.

reasons for promoting zooplankton production during the prestocking period. Alone, either one would justify a very high priority for zooplankton production. Together, they make proper zooplankton management and production the top priority in striped bass culture.

The young of many fish are sensitive to sunlight (Breder 1962). Rees and Cook (1982) have demonstrated detrimental effects of direct sunlight on 6 to 8-day old striped bass larvae. For this reason, culture ponds should always be stocked at night or at dusk. Many culturists hold and transport their larvae in dark containers, even at night, and dye their ponds or partially cover them with black plastic sheets to provide additional shelter.

Dense phytoplankton blooms apparently provide adequate shelter for the fish at Marion. Phytoplankton production is given priority over sunlight protection. Dyes and shelters inhibit photosynthesis; therefore they are used only as a last resort. When the plankton bloom is lost, however, nigrosine dye at 1 mg/L is applied to the pond. The nigrosine will dissipate after a few days, allowing the culturist to reestablish his phytoplankton bloom.

Stocking Guidelines

Guidelines were developed to implement the previously mentioned three-point program. These guidelines require that no pond will be stocked unless:

(1) secchi disc visibility is less than 75 cm (50-60 cm preferred)

(2) at least 100 crustaceans per liter are present, including nauplii larvae and/or medium sized cladocerans such as <u>Ceriodaphnia</u> (several hundred to approximately one thousand preferred)

(3) average pond temperature exceeds 18°C and all other water quality parameters are within acceptable range

(4) when pond temperatures are less than 23°C, the fish are stocked at dusk. When afternoon temperatures exceed 23°C, stocking is

accomplished no later than one hour prior to sunrise.

No strict guideline determines stocking rate. Preferred stocking rates at other hatcheries in the southeast range from 110,000 to 2,200,000 per hectare, with the concensus being 200,000 to 400,000 per hectare. Optimum stocking rate at Marion is 300,000 to 400,000 per hectare. Stocking rates are adjusted up or down depending on the pond's productivity.

When stocking rates are high, proper fertilization and zooplankton management becomes critical. Inexperienced culturists are therefore advised to stock at low rates (100,000 to 200,000 per hectare) because low rates leave more room for error than do high rates. Experience will eventually allow those culturists to identify stocking rates which are best suited for their particular hatcheries.

There is no general agreement on how long larvae should be held prior to stocking. At some hatcheries the larvae are stocked when they begin to feed (usually 4 to 7 days after hatching) and at other hatcheries the fish are held an additional five or six days on an *Artemia* diet. The former method does not require an *Artemia* program and thus saves much time and expense; the latter produces larger and older larvae which are apparently less fragile than younger larvae. Also, holding the larvae on *Artemia* provides additional flexibility for the culturist. There is no obligation to stock the ponds when the fish are 5 days old - they may be held until pond conditions are proper. At Marion, stocking is predicated upon pond conditions rather than the age of the larvae, thus they are held on *Artemia* until they are 7 to 12 days old. It is not advisable to stock larvae which have not yet begun to feed, to hold fry older than 5 days posthatching without feeding them, or longer than 12 days on an *Artemia* diet.

Survival after stocking may be confirmed by several different methods, none of which is foolproof. Some culturists attract the larvae with a bright light, while others depend on visual confirmation of survival in cages. At Marion, survival is confirmed with a 1mm mesh tow net ($0.5m^2$ mouth opening). If survival cannot be confirmed and additional larvae are available, the pond is usually drained, refilled, fertilized, and restocked.

POND MAINTENANCE

After the ponds are stocked, the culturist's role is essentially that of a caretaker. The primary objective is to prevent or remedy adverse culture conditions. Any number of such conditions may arise, but most will fall into one or more of the categories listed below:

(1) insufficient food supplies
(2) inadequate water quality
(3) overgrowth of nuisance vegetation
(4) outbreak of fish disease

Maintenance of Food Supplies

At Marion, striped bass production is enhanced when the fish are provided with high-quality zooplankton forage during the latter part of culture period - a task which is easier said than done. The tremendous predatory pressure exerted by the fish makes it very difficult for the choice zooplankters to maintain their numbers more than 1 to 2 weeks poststocking (Turner 1981b). In order for them to do so they must: (1) become abundant before the predatory pressure is exerted, and (2) maintain maximum growth and reproduction rates. Translated into management terms, the culturist must: (1) establish large populations of high-quality zooplankton prior to stocking and (2) maintain high fertilization rates as permitted by water quality.

Nevertheless, in many instances zooplankton is depleted prematurely. When this occurs, the fish convert from zooplankton to an insect diet (Turner 1982). At Marion, this diet is apparently inadequate to support large numbers of fish, for such ponds are characteristically low in production. There are options available to the culturist following premature zooplankton depletion, but they are seldom completely successful. These options include: supplemental feeding with dry food, supplemental zooplankton inoculation, harvest and redistribution among ponds, and harvest and distribution to final destination.

Perhaps the most widely practiced method is supplemental feeding. After the zooplankton is depleted, artificial feed such as Silver Cup Salmon Diet can be offered at a rate of 5 kg/ha/day. Feeding rates are gradually increased to 20 kg/ha/day. The feed should be offered 2 to 6 times each day (Bonn et al. 1976).

Many hatcheries incorporate supplemental feeding into their normal routine regardless of zooplankton conditions. Benefits of routine feeding have not been accurately measured, so at this time it is not possible to make a definite recommendation either for or against it. Routine feeding is not practiced at Marion because of its manpower requirements and because its benefits are yet undefined. This policy could very well be changed, however, as additional information becomes available. For the time being, each culturist must decide whether routine feeding is of benefit at each hatchery.

Supplemental inoculation of zooplankton after stocking holds some promise in alleviating food shortages. However, there are particular logistical problems which limit its usefulness. The problems arise from the enormous quantity of zooplankton usually necessary to reverse impending zooplankton depletion. It is very difficult to locate a source for vast quantities of zooplankton and, if located, it is very difficult to transport them from one location to another. At Marion, supplemental inoculation seems useful when catfish production ponds (usually a good zooplankton source) are adjacent to striped bass ponds with failing zooplankton populations. In this situation, screened water can be pumped from the former into the latter and, over a period of several days, zooplankton conditions usually improve. When no immediate zooplankton source is available, supplemental inoculation is logistically impossible.

Fish are occasionally harvested at Marion which are smaller than minimum size standards. If pond space is available, these fish are sometimes restocked for additonal growth. In limited application, this technique may be surprisingly useful. At Marion, production rates in excess of 10 kg/ha/day over a 2 week period have been observed in ponds recieving such fish. This provides the culturist with another option when pond zooplankton fails — the fish may be harvested and redistributed among available ponds. This technique has obvious disadvantages. Fishless ponds must be available and double-harvest entails extra work. In some instances, however, this approach may prove useful.

The best treatment for premature zooplankton depletion is to prevent it from occuring. Again, as with weed control, an ounce of prevention is worth a pound of cure. For this reason, much more effort is delagated to the prevention of zooplankton depletion at Marion than to corrective measures following depletion.

Maintenance of Water Quality

At Marion, water quality maintenance is not assigned as high a priority as food production either before or after stocking. This does not mean that water quality is not important. Most culturists already know (usually from unfortunate experience) that an unlimited food supply is of little value when there is no dissolved oxygen. Oxygen is the focal point in water quality maintenance. Also of concern are pH, temperature, carbon dioxide, and nitrogen.

Dissolved oxygen should be maintained above 3 to 4 mg/L. When oxygen declines below that point the culturist should consider supplemental aeration. However, supplemental aeration should be avoided whenever possible (Bonn et al. 1976).

The culturist's most valuable asset in maintaining dissolved oxygen is a large and healthy phytoplankton bloom. Abundant phytoplankton is virtually the only source capable of supplying enough oxygen to support the aggresive organic fertilizer program employed at Marion. Oxygen depletion seldom occurs at Marion,even in the most heavily fertilized ponds, providing that the phytoplankton population remains healthy.

Occasionally, however, phytoplankton populations will suddenly, and unexpectedly, collapse. Phytoplankton collapse may be associated with sudden cold and cloudy weather, cold rain, or other abrupt climatic changes. Observations at Marion indicate that sudden collapse may also be associated with the preflooding treatment of simazine at 10 kg/ha or less and, for this reason, minimum application rates are recommended for preflooding.

Pond-water pH should remain between 7.0 and 9.5. Low pH seldom occurs when phytoplankton is abundant and is therefore not discussed here. High pH is usually associated with lush growths of aquatic weeds or very dense phytoplankton blooms. High pH is another often-quoted reason for avoiding inorganic fertilizers. While this may be a valid point of concern at other hatcheries, the use of inorganic fertilizers seldom leads to pH problems at Marion. This is probably due to the liberal use of organic fertilizers (which produce acid, thereby lowering pH), and an annual or semiannual application of agricultural limestone of 1000 kg/ha.

If extremely high pH occurs, it can be remedied by flushing out excess plankton with fresh water or by partial kill of phytoplankton with chemical treatment. Boyd (1979) describes an emergency treatment with alum which provides temporary relief from high pH conditions.

High fertilizer rates may also lead to high levels of toxic ammonia, nitrites, or carbon dioxide. All have been implicated in fish kills or in reduced fish production (Boyd 1979). Striped bass culture ponds frequently develop sublethal concentrations of these wastes, but existing data (Geiger 1981; Turner 1982) indicate that lethal concentrations are unlikely to occur in ponds fertilized as described in this report.

Regardless of the fertilization regime which is adopted, or the care which is lavished on the ponds as they are prepared for stocking, occasional water quality failure is virtually inevitable. Oxygen depletion is probably the most common and most serious failure. Due to the density of fingerlings (often in excess of 250,000/ha) and the high monetary value of the fish ($.30 to $.50 per fish), the consequences of even a single acute failure can be enormous. In 1981 an intensive water quality monitoring and maintenance program was instituted at Marion. The objective of this program is to identify water quality failures before they occur, or to mitigate their effects as soon as possible after their occurrence. Since the implementation of this program, not a single pond has been lost due to water quality failure (including largemouth and smallmouth bass, sauger, channel catfish, and any other species on station when the program is in effect). For this reason, regular monitoring of water quality is highly recommended to the beginning culturist.

The program is built upon: (1) the capability to accurately measure critical water quality parameters in less than 2 minutes per site, and (2) the delegation to one specific individual, as his primary responsibility, the task of making those measurements twice each day. The first requirement is necessary at Marion because of the large number of ponds (in excess of 60 ponds/day), and may be met with a polarographic oxygen meter. At other hatcheries where fewer ponds are involved, a simple test kit would be sufficient. The second requirement adds the key ingredient to the program — consistency, regularity, and dependability.

With such a program in place, culturists have reliable and up-to-date information at their fingertips. Simple comparison will allow them to keep track of day-to-day trends and identify chronic problems. With experience, culturists will often be able to anticipate acute oxygen failures before they occur. When acute failure cannot be anticipated, the routine monitoring program will place a hatcheryman at the scene in time to take action to prevent a fishkill.

Timely detection is, however, only the first step. After a failure has been detected a culturist must have some capability to combat it. At Marion, electricity is provided to each pond and mechanical aeration is possible. Although aerators do not prevent conditions of low oxygen, they will usually supply enough oxygen to prevent a severe fish kill.

Other methods for combating oxygen depletion are available. Potassium permanganate is frequently recommended for this purpose (Bonn et al. 1976), but Tucker and Boyd (1977) question the value of this treatment. Based upon their recommendations, potassium permanganate should be used with caution, and variable results should be expected. Flushing the pond with fresh water is usually beneficial and application of inorganic fertilizers occasionally stimulates photosynthesis. All of these methods are more useful against chronic low-oxygen conditions than against sudden, acute failure. Acute failure usually requires immediate vigorous response such as with mechanical aeration.

Fish Diseases

Parasites and diseases of striped bass in culture systems, and treatments for those parasites and diseases, are discussed elsewhere in this conference and by Bonn et al. (1976). It is beyond the scope of this report to cover this aspect of culture in detail, so the reader is referred to those works for specific information. It should suffice here to say that the best treatment of any disease is prevention, and that the role of proper nutrition in disease prevention should be emphasized. Striped bass that are properly nourished are generally healthy and robust, whereas those suffering from food shortages are usually in poor health. Water quality maintenance and the avoidance of stress are also important in disease prevention.

HARVEST AND DISTRIBUTION

"Guidelines for Striped Bass Culture" opens it's "Harvest" section with a warning about acute shock of fingerlings during harvest procedures. This warning is not to be taken lightly, because striped bass fingerlings are extremely vulnerable to stress and shock during harvest. Striped bass do not tolerate heat as well as most other warm-water fishes (Coutant and Carroll 1980) and their vulnerability to shock is apparently increased by the elevated water temperature which normally accompanies pond harvest. For this reason, all possible precautions are taken to reduce stress and shock during the harvest and distribution procedures. The fish are handled as little as possible, direct sunlight is avoided, the water temperature is kept as cool as possible, aeration is provided when needed, and the fish are never handled and then placed directly into fresh water. Instead, they are placed into 0.5% to 1% salt solution and then gradually acclimated to fresh water.

The duration of the culture period is largely determined by hatchery policy. At Marion the culture system is geared towards production of fingerlings ranging in weight from 0.3g to 1.0g (1,000 to 3,000 fish/kg). This requires a culture period of 25 to 35 days, however, the culture period at other hatcheries may last much longer. Near the end of the culture season the culturist should check the growth of the fish every week so that the ponds can be drained as soon as the fish are large enough to meet stocking requirements.

The culturist should check his ponds for nuisance insects one week prior to draining. Diesel fuel is used to eliminate air-breathing insects. Dragonfly nymphs or fairy shrimp may be eliminated with Baytex or Dylox at 0.25 mg/L. Treatment should be made several days prior to harvest to allow time for the dead animals to decompose.

Timing is important when draining ponds. Drawdown should be timed so that sufficient water is left in the pond to protect the fish during the heat of the day. Ponds should never be drawn down in the morning and left throughout the day because the water temperature may exceed safe limits. Instead, ponds should be drawn down late in the day so that temperatures do not rise any higher than necessary. As a rule of thumb, healthy striped bass fingerlings tolerate temperatures to 32°C if they are handled carefully; above that temperature the culturist must be very careful. Temperatures of 35°C or higher may be fatal regardless of the culturist's precautions.

Glass V-traps are highly recommended for harvesting striped bass. Techniques described by Bonn et al. (1976) are employed at Marion. The trap is placed in front of the drain screen. As the water flows out of the pond, the fish congregate around the screen and enter the trap. As the fish accumulate in the trap, they are dipped out with a knotless net and immediately placed in a 60L tub of oxygenated water containing 1% to 2% salt (uniodized). They are promptly carried to a holding facility where the tub is submersed in a vat of 1% salt solution. V-trap harvest is preferred at Marion because handling stress is minimized.

When the ponds reach a depth of 15 to 20 cm the drain valve is closed, but the V-trap can still be used. The traps are left in the catch basin and a 1/3 horsepower agitator is placed so that a stream of water flows through the trap. Lights may be set up around the basin to attract the fish. Once the fish enter the basin the current directs them into the trap. V-traps operated in this manner will catch fish indefinitely, and do not depend on outflow as described in "Guidelines for Striped Bass Culture." The traps are removed when they are no longer effective, but the agitators are left in place throughout the night to prevent oxygen depletion.

V-traps may remove up to 80% of the fish (Braschler 1974; Bonn et al. 1976), but more commonly only 40% to 60% are removed. Occasionally very few fish are trapped and the pond must be drained to harvest them. Reasons for trap failure are not clear, but efficiency appears to decline as temperature increases, as the size of the fish increases, and as the size difference among the fish increases.

Draining is accomplished early in the morning before the sun overheats the remaining pond water. The fish are drawn into the catch basin, captured with seines, and placed in an oxygenated 1% salt solution for transfer to the holding facility. Fish harvested in this fashion are subjected to more stress than with V-traps and are more likely to suffer acute shock. For this reason the culturist must be extremely cautious during the final stages of harvest.

Because Marion ponds are relatively small (0.3 to 0.6 ha), the entire harvest operation for each pond can be completed within 24 hours. Pond draining begins early in the morning, drawndown and V-trap operations begin in late afternoon, and final draining is completed early the following morning. At hatcheries where ponds are larger or draining is slower, the most effective schedule may differ from that described above. Nevertheless, basic principles of striped bass pond draining still apply. All ponds must be drained so that the fingerlings receive maximum protection from solar radiation and resulting elevated temperatures; all precautions must be taken to prevent excessive handling stress and acute shock.

When the fish arrive at the holding facility, they are placed in vats containing 1% salt solution. Calcium chloride may also be added at a rate of 75 mg/L to relieve stress. A flow-through antibiotic treatment of 100 mg/L (commercial product) of furacin (5% active ingredient) is also administered. Salt and furacin are replenished as required. After the fish are harvested, placed in the vats, and treated, they should not be disturbed for at least 24 hours (Bonn et al. 1976).

The fish are delivered the day after draining. Striped bass should not be transported in freshwater. At Marion, the hauling media consists of 1% salt solution, 75 mg/L calcium chloride, 100 mg/L furacin, and 5 ml antifoam surfactant. No anaesthetic is used at Marion, but many other agencies add a tranquilizing dose of quinaldine at 0.25 mg/L. Critical water quality parameters are measured in the hauling tank before the fish are loaded.

Before the fish are handled the vats are once again salted to 1%. A subsample (or a series of subsamples, depending upon accuracy required) is collected to estimate the number/weight ratio. The fish are then weighed onto the truck, taking care to keep them immersed in a salt solution at every possible opportunity. Hauling density is variable, but the fish/water ratio never exceeds 60 g/L. On long deliveries the ratio never exceeds 30 g/L.

Delivery trucks are equipped with agitators and compressed oxygen. All hauling tanks are insulated to minimize rise in temperature. Tanks are fitted with quick-release, sliding gate valves. Deliverymen are required to periodically check equipment and to measure water quality parameters en route. Ice may be added to reduce temperature as required. Upon arrival at the stocking site, water quality is immediately checked and the tempering process is begun. Water is pumped from the lake into the tank with a 2 horsepower gasoline-engine pump. Tempering requires 1 to 2 hours with a 1400 L tank. The fish are released when temperature and conductivity inside the tank matches that of the receiving waters.

A 3 m length of bell-end PVC pipe is placed so that the bell-end fits over the quick release opening and the straight end is in the water. The sliding gate is opened and the fish travel through the pipe into the water.

Fish must be dipped from tanks without the quick release valve. The dip method is considered inferior to quick release because the extra handling may produce injury and because the fish are released into freshwater after handling, which may result in fatal, acute shock.

REFERENCES

ALABAMA FISHERIES SECTION, DEPARTMENT OF CONSERVATION. Undated, Fisheries leaflet no. 15, A guide for fertilizing ponds with liquid fertilizer. Montgomery, Alabama.

ALLAN, J. D. 1976. Life history patterns in zooplankton. Amer. Naturalist 110:165-180.

BONN, E. W., W. M. BAILEY, J. D. BAYLESS, K. E. ERIKSON, and R. E. STEVENS, eds. 1976. Guidelines for striped bass culture. Striped Bass Committee, So. Div., Am. Fish. Soc. 103pp.

BOYD, C. E. 1981a. Soluability of granular inorganic fertilizers for fish ponds. Trans. Am. Fish. Soc. 110(3):451-454.

_____1981b. Comparison of five fertilization programs for fish ponds. Trans. Am. Fish. Soc. 110(4):541-545.

_____1979. Water quality in warmwater fish ponds. Auburn Univ., Agric. Exp. St., Auburn, AL 359pp.

BREDER, C. M., Jr. 1962. On the significance of transparency in osteichthid fish eggs and larvae. Copeia 1962:561-567.

BRASCHLER, E. W. 1974. Development of pond culture techniques for striped bass (*Morone saxatilis*). Proc. 28th Ann. Conf. S.E. Game Fish Comm. 28:44-48.

COUTANT, C. C., and D. S. CARROLL. 1980. Temperatures occupied by ten ultrasonic-tagged striped bass in freshwater lakes. Trans. Am. Fish. Soc. 109(2): 195-202.

DAVIDSON, R. G. and C. E. BOYD. 1981. Phytoplankton response to liquid fertilizers. Prog. Fish-Cult. 43(3):126-129.

EMONDSON, W. T., G. W. COMITA, and G. C. ANDERSON. 1962. Reproductive rate of copepods in nature and its relation to phytoplankton production. Ecology 43:625-634.

ELDRIDGE, M. B., J. A. WHIPPLE, DANA ENG, M. J. BOWERS, AND B. M. JARVIS. 1981. Effects of food and feeding factors on laboratory-reared striped bass larvae. Trans. Am. Fish. Soc. 110:111-120.

GEIGER, J. G. 1981. Zooplankton dynamics in intensive pond culture of striped bass. Unpubl. research, U. S. Fish Wildlife Serv., S. E. Fish Cult. Lab., Marion, AL.

HARPER, J. L., and R. JARMAN. 1971. Investigation of striped bass, _Roccus saxatilis_, culture in Oklahoma. Proc. 25th Ann. Conf. S.E. Game Fish Comm. 25:510-512.

HUMPHRIES, E. T., and K. B. CUMMING. 1973. An evaluation of striped bass fingerling culture. Trans. Am. Fish. Soc. 102(1):13-20.

HUNTER, J. R. 1981. Feeding ecology and predation of marine fish larvae. In, Marine fish larvae, Reuben Lasker, ed. University of Washington Press, Seattle and London. 131pp.

HURLBERT, S. H., M. S. MULLA, and H. R. WILLSON. 1972. Effects of an organophosphorus insecticide on the phytoplankton, zooplankton and insect populations of freshwater ponds. Ecol. Monographs 42(3):269-299.

LOGAN, H. J. 1967. Comparison of growth and survival rates of stripes bass and striped bass X white bass hybrids under controlled environments. Proc. 21st Ann. Conf. S.E. Game Fish Comm. 21:260-263.

McCAULEY, E., and J. KALFF. 1981. Empirical relationships between phytoplankton and zooplankton biomass in lakes. Can. J. Fish. Aq. Sci. 38:458-463.

McCRAREN, J. P., and T. R. PHILLIPS. 1977. Effects of Masoten (Dylox) on plankton in earthen ponds. Proc. 31st Ann. Conf. S.E. Assoc. Fish and Wildlife Agencies 31:441-448.

METZGER, R. J., and C. E. BOYD. 1980. Liquid ammonium polyphosphate as a fish pond fertilizer. Trans. Am. Fish. Soc. 109(5):469-586.

REES, R. A. 1978. Improvement of striped bass fingerling production. Georgia Dept. Natural Res., Game and Fish Div., Atlanta, GA. 37pp.

REES, R. A., and S. F. COOK. 1982. The effects of sunlight intensity on the survival of stripe bass X white bass fry. Proc. 37th Ann. Conf. S.E. Assoc. Fish Wildlife Agencies. 37:in press.

SANDOZ, O., and K. J. JOHNSTON. 1965. Culture of striped bass. Proc. 19th Ann. Conf. S.E. Game Fish Comm. 19:390-394.

SCRUGGS, G. C. Jr., and J. C. FULLER, Jr. 1954. Indications of a freshwater population of striped bass, *Roccus saxatilis*, in Santee-Cooper Reservoir. Proc. S.E. Asso. Game Fish Comm. 8:64-70.

SNOW, J. R. 1977. A technique for controlling weeds in striped bass rearing ponds. Proc. 31st Ann. Conf. S.E. Fish Wildlife Agencies. 31:322-328.

STEVENS, R. E. 1966. Hormone induced spawning of striped bass for reservoir stocking. Prog. Fish Cult. 28:19-28.

TUCKER, C. S., and C. E. BOYD. 1977. Relationships between potassium permanganate treatment and water quality. Trans. Am. Fish. Soc. 106:481-488.

TURNER, C. J. 1982. Investigations into striped bass production at Marion Fish Hatchery, FY 82 Annual Report Dept. Conservation and Nat. Res. 22pp.

_____1981a. Investigations into striped bass production at Marion Fish Hatchery, FY 81 Annual Report, Dept. Conservation and Nat. Res. 15pp.

_____1981b. Zooplankton of striped bass culture ponds of six southeastern hatcheries. Unpubl. rept. to Striped Bass Committee, So. Div., Am. Fish. Soc. 12pp.

_____1980. Evaluation of three organic fertilizers in shaded and non-shaded striped bass ponds. FY 80 Annual Report, Dept. Conservation and Nat. Res. 23pp.

VON WINDEGUTH, D. L., and R. S. PATTERSON. 1966. The effects of two organic phosphate insecticides on segments of the aquatic biota. Mosquito New 26:377-380.

WRIGHT, J. C. 1965. The population dynamics and production of *Daphnia* in Canyon Ferry Reservoir, Montana. Limn. Oceanography 10:583-590.

Observations on Pond Production of Hybrid Bass Fingerlings

James W. Kahrs

On June 10, 1981, 40,000 hybrid bass fingerlings were stocked into a 1-acre pond at our Osage Beach facility. The fingerlings were 38 to 50 mm in length and weighed approximately 1.6 grams. A flow of 15 gpm of aerated spring water flowed through the pond until harvest.

The fish were fed twice daily using a mixture of crumbled floating trout feed (38% protein). The fish fed voraciously at all water levels from the surface to the bottom. When the electric golf cart or pick-up truck used in feeding operations came near the pond the fish would rapidly congregate in the area where they were normally fed.

The fish were sampled throughout the summer and were sold when necessary to meet commitments.

On August 17th, 7,000 fingerlings were harvested for delivery to customers. These fish averaged 127 mm total length (TL) but no weights were taken. Survival was high and no disease problems were encountered. During the last week in August 1981 a six-inch rain occurred during a 2-hour period. The creek that lies adjacent to the fingerling pond rose to the highest level I had seen in nearly 30 years and completely flooded the pond containing the hybrid bass. Some fish were recovered from the original pond and adjacent ponds, but many made their way to the Lake of the Ozarks which lies less than one-quarter mile downstream.

In June 1982, the same pond was stocked with 22,000 fingerling hybrid bass 25 to 50 mm TL. Periodic harvests of some fish were made until the pond was completely harvested in early September 1982. Survival rates were not estimated and lengths and weights were not taken, but lengths and weights were in the range of the previous year.

I am extremely impressed with the hybrid bass. I feel that it has great potential as a food fish as well as a recreational fish.

Intensive Culture of Striped Bass: A Review of Recent Technological Developments

James M. Carlberg, Jon C. Van Olst,
Michael J. Massingill and Timothy A. Hovanec

The artificial propagation of striped bass, Morone saxatilis, was first attempted in 1874 at the Weldon Hatchery in North Carolina. Hatchery-reared fry were released for enhancement of sport and commerical fisheries. It was not until the formation of the Santee Cooper Reservoir in 1941, and the subsequent establishment of a reproducing population, that additional hatcheries were constructed (Stevens 1979). Successful establishment of striped bass in landlocked lakes by the impoundment of wild fingerlings, together with the lack of success in stocking wild caught adults and hatchery-reared fry, caused a demand for large numbers of fingerlings for transplantation programs. As a result, the Edenton National Fish Hatchery in North Carolina and the Moncks Corner Hatchery in South Carolina were constructed in the early 1960s (Kerby 1982). Techniques for the production of fingerlings by pond culture methods were developed at these hatcheries (Bayless 1972; Braschler 1974; Bonn 1976). Refinement of controlled spawning techniques soon followed, using hormonally induced ovulation and intraovarian catheterization (Stevens 1966). Striped bass have now been successfully introduced into numerous inland lakes and reservoirs (Bailey 1974).

Conventional extensive pond culture methods allow only minimal control over critical culture parameters. Conversely, high density culture systems, with limited water volumes, allow for the control of the culture environment. Other advantages are the accessibility and ease of observation, low maintenance and reduced handling in grading and harvesting procedures. Disadvantages are the sensitivity to rapid fluctuations in water quality and outbreaks of stress related disease, the dependence on mechanical devices, a need to supply all the nutritional requirements in the absence of natural foods, the need for alarm and backup systems and a requirement for experienced personnel. Nevertheless, the added risk is often justified by the increase in production capacity, where harvest densities can be three levels of magnitude greater than traditional pond production levels.

The striped bass is an excellent species for intensive culture. It is adaptable to controlled environments, exhibits rapid growth rates in captivity, accepts artificial feeds, has broad physiological tolerances and has a high market value. Intensive culture methods have been applied to all developmental stages of striped bass, from larvae to mature adult. Historically, eggs have been incubated in McDonald hatching jars and resulting larvae and fry reared in aquaria and troughs (Bayless 1972; Bonn 1976). Traditional pond culture methods are still commonly used during the nursery period when fry are reared to fingerling. However, in recent years a variety of intensive culture systems have been tested for rearing fingerlings and for the production of advanced fingerlings and food fish under controlled culture conditions. These intensice culture systems include floating cages, tanks, raceways, and silos. The following is a review of some of the significant accomplishments made in the continued effort to develop intensive culture techniques for striped bass.

INTENSIVE CULTURE OF FRY

Dennis Wildlife Center,
South Carolina Wildlife and
Marine Resources Department,
Bonneau, South Carolina
(Bayless 1972; Bayless and Harrell pers. comm.)

In the late 1960s striped bass culture techniques were refined at the Moncks Corner Hatchery. The culture sequence was modified to include a 7- to 15-day period of rearing fry in 303-l fiberglass coated wood troughs prior to stocking in ponds. Fry were fed brine shrimp nauplii as an initial food source to increase their strength and development prior to their release in ponds. This method allowed for some control during the critical first few days of feeding, and has often been found to increase the subsequent survival of striped bass reared in ponds. Troughs are now traditionally used to hold fry at several hatcheries prior to pond stocking. Fry are often stocked in troughs at 200 fry/l and fed trout starter mash and ground fish at 10% of their body weight per day.

Fish Culture Development Center, U.S. Fish and Wildlife Service, Tishomingo. Oklahoma (Inslee 1977)

Between 1972 and 1974, investigators in Oklahoma tested the use of 1-m3 cages floating in tanks and ponds as an alternative to the use of aquaria for rearing larvae until swim-up. The cages were constructed of wood frames covered with saran screen. Following 5 days of culture, the cages were submerged and the fry were allowed to escape into the production ponds. In 1974 cages were suspended in tanks equipped with wooden paddle agitators and after 3 years of study the average survival of fingerlings reared in ponds stocked with fry that were initially held in saran cages was 35.7%, almost twice the survival rate of aquaria-reared fry (18%) stocked in ponds. Problems with this method of larval rearing were the difficulty of stocking the cages and the lack of control over light and other environmental conditions. Nevertheless, it was demonstrated that larvae cultured in cages prior to pond release required less care and manpower than aquaria methods.

Texas Parks and Wildlife Department,
Fisheries Research Station,
Palacios, Texas
(Colura et al. 1976; Hysmith pers. comm.)

Research was conducted on the culture of striped bass larvae by intensive methods involving the use of circular tanks. Larvae were reared from day 3 to day 10 in 0.4-, 2.5-, or 6.8-m3 tanks. The tanks were operated as a closed system, at a temperature of 19°C and a salinity of 6-8 ppt. Prior to stocking in brackish water ponds fry were fed brine shrimp nauplii. This study showed that fry could be produced by intensive tank culture methods in a closed system at low salinities.

INTENSIVE CULTURE OF FINGERLINGS

The rearing of fry to 2.5- to 5.0-cm fingerlings traditionally involves stocking 5-day-old fry produced in aquaria into earthen ponds of about 0.5 ha in size, at densities of 250,000/ha. They feed on zooplankton populations which are enhanced by fertilization and supplemented with artificial feeds (Bonn 1976). Survival of fry grown by extensive pond culture methods is usually less than 20% at harvest after 30-50 days. Over the past 20 years considerable effort has been made to improve survival rates during this critical stage of development. These efforts include several attempts to utilize intensive culture techniques.

North Carolina State Fish Hatchery
Weldon, North Carolina
(Tatum et al. 1965).

The first recorded efforts to culture fingerlings by intensive methods were conducted at Weldon in the early 1960s. Culture experiments were conducted in 20-m3 concrete pools stocked with 2-day-old larvae at densities of 0.5/l and 2.5/l. They were fed live zooplankton for 3 weeks and trout starter meal after the second week. Sixteen weeks of culture resulted in the production of 6- to 13-cm fingerlings with an

85% survival rate. This study showed that fry could be cultured successfully in outdoor concrete pools and would readily consume artificial feed.

Edenton National Fish Hatchery
U.S. Fish and Wildlife Service,
Edenton, North Carolina
(Anderson 1966; Wirtanen and Ray 1971;
Atstupenas pers. comm.)

Since the early 1960s, the Edenton Hatchery has been one of the primary sites for the production of fingerlings in ponds. The first recorded attempt to rear fingerlings for enhancement purposes, as opposed to stocking sac-fry, was at the Edenton Hatchery in 1964 (Anderson 1966). Sac-fry obtained from the Weldon Hatchery were grown in 450 l troughs supplied with 21°C aerated well water at a rate of 5.7 lpm. The fry were fed emulsified shrimp ten times a day. After 15 days in the troughs mortality in the first group was greater than 75%. It was concluded that emulsified shrimp was not an adaquate diet for an extended culture period.

A second group of fry were grown in the troughs for only 5 days prior to stocking in a pond. Subsequently, 30,000 fingerlings weighing 4-5 g each were harvested from the pond after 83 days. From this study researchers concluded that stocking ponds with older fry resulted in greater survival to fingerling size and that providing the appropriate feed was the most critical factor.

Routine operations at the Edenton Hatchery involve stocking fry in late May, and 5-cm fingerlings weighing 0.45 g are harvested after 35 days in early July. Some fingerlings are graded and restocked for production of advanced fingerlings and others grown to maturity in an effort to develop a captive broodstock.

Despite the relatively successful pond culture operations at Edenton some research there has involved the development of alternative culture techniques to increase survival rates. In 1964 0.6-g fry were stocked in indoor concrete holding tanks at a density of 4 g/l. The tanks were supplied with 18°C well water at a rate that produced three water changes per hour. The fry were fed commercial trout feed at a rate of 5% body weight per day. There were considerable disease problems caused by *Flexibacter columnaris*, attributed to the relatively high stocking density.

Other studies on the intensive culture of fingerlings were conducted in two 45.4 m3 outdoor fiberglass raceways. Fry weighing 1.4 g were stocked at two densities, 0.5 g/l and 1.0 g/l. Both raceways were supplied with 24°C well water at a rate of 208 lpm. The first raceway received 13,810 fry and yielded 5,715 fingerlings at 7.5 g each after 140 days (41.4% survival). The second raceway received 29,795 fry and yielded 10,159 fingerlings at 11.1 g each after 187 days (34.1% survival). The results of the raceway experiment showed that fingerlings could be cultured in intensive systems from a relatively high stocking density of 2.76 kg/m3 with a survival percentage nearly twice that achieved in ponds. In addition, this research revealed the advantage of raceways for the control of predation and treatment of disease.

Lower Fisheries Research Unit
Department of Fisheries and Applied Aquaculture
Auburn University, Auburn, Alabama
(Kelley 1967; Shell 1972;
Snow 1977, 1979; Braid 1981)

A study similar to the one conducted at Edenton was performed at the Agricultural Experiment Station of Auburn University in 1967 (Kelley 1967). Fourteen 98-l stainless steel troughs were each stocked with 100 15- to 35-mm fry. Each trough had a flow of 2.6 lpm of filtered water and fry were fed floating Purina Trout Chow and beef liver at 10% body weight per day. After 16 weeks of culture, fingerlings with an average size of 70.6 mm were produced with only 14.7% mortality. The length attained by a similar group reared in ponds was only 54.9 mm. The results from this study showed that fingerlings could be reared in intensive culture systems when fed an adequate amount of appropriate feed.

Similar studies conducted at Auburn University from 1971-1973 assessed the value of extended culture of larvae in McDonald hatching jars prior to stocking fry in troughs for intensive production of fingerlings (Shell 1972). Larvae were stocked in 7.6-l and 22.7-l jars at densities of 10,000 and 30,000, respectively. There were considerable problems with infestation by fungus that resulted in total mortality in the 22.7-l jars and only 0.02% survival in the smaller 7.6-l jars. In subsequent attempts 29-day-old fry were stocked into aluminum troughs that received 17.8°C stream water adjusted to a salinity of 5 ppt. The fry were fed brine shrimp nauplii every 3 hours and dry feed every 15 minutes. Again the results were discouraging, with a mortality of 97% resulting from cannibalism and unknown causes. At that time it was concluded that delayed stocking of fry increased survival rates to fingerling size, however due to the high mortality rates of fry in hatching jars and troughs, this proved to be of little advantage.

The importance of diet was apparent from the first two studies. Therefore, in 1974 a series of experiments were conducted to determine a suitable diet for fry in intensive culture. Several combinations of brine shrimp with commercial trout feeds were tested (Braid 1981). Later studies at Auburn University in 1976 and 1977 investigated alternative methods for the intensive culture of fry (Snow 1977 and 1979). In 1976 a closed recirculated culture system was tested. Although this preliminary trial was successful, when the fry were stocked in brackish water ponds at the Claude Peteet Mariculture Center, Gulf Shores, Alabama, more than half the resulting fingerlings showed signs of scoliosis. This was attributed to possible vitamin deficiencies in the feed.

Continuing work in 1976 concerned the evaluation of tank culture systems. In one experiment five 2.4-m3 circular plastic pools were supplied with water at 25°C that was pre-treated with organic or inorganic fertilizer to stimulate the growth of natural food organisms. Attempts to stock 4-day-old fry resulted in total mortality. However, when 16-day-old fry were stocked, 23.1% survived to fingerling stage, compared to 21.2% for similar groups cultured in ponds. This study indicated that tank culture methods were suitable for advanced fry.

An additional study conducted in 1976 tested 853-l cages constructed of saran screen and suspended in 1.5-m3 concrete tanks. Fry were fed brine shrimp nauplii five times per day. After 15 days, survival was only 22.6% due to high rates of cannibalism. It was believed that brine shrimp nauplii might not provide all the nutritional requirements of older fry. In 1977 investigations were initiated to identify an alternative to brine shrimp nauplii, since this feed was in short supply and relatively expensive. Again, saran cages were used in the 1.5-m3 tanks. Two sizes of cages were tested, with volumes of 0.5 and 7.5 m3. A third system consisted of two circular fiberglass tanks, 1.5 and 2.4 m in diameter and volumes of 1.3 and 3.4 m3, respectively. The study involved the testing of several artificial dry feeds in combination with natural food organisms grown in adjacent ponds. The results demonstrated that zooplankton could be cultured in ponds and transferred to intensive fry culture systems as needed.

Virginia Institute of Marine Sciences
Gloucester Point, Virginia
(Kerby and Joseph 1978)

Concurrent with the tank culture studies conducted at Auburn University, similar systems were being investigated at the VIMS Laboratory. Initial attempts at tank culture of 900,000 larvae resulted in 4,000 fingerlings, giving only a 0.4% survival. Subsequent studies conducted in 1970 were more successful. Two series of experiments were conducted to compare the growth and survival of striped bass and striped bass x white perch hybrids cultured in intensive systems. Two 1,300-l fiberglass tanks were supplied with 24°C well water at a rate of four exchanges per day. Fry were fed Purina Trout Chow twice per day. The first trials were conducted for 332 days, yielding 203-g striped bass fingerlings with a survival of 46%, and 171-g hybrids with 82% survival. The second trials were conducted for 331 days, during which time the striped bass attained an average size of 174 g with 38% survival, and the hybrids reached 196 g with 87% survival. This study demonstrated that hybrids were clearly more hardy than striped bass. In addition, it was determined that hybrids could be grown to a harvest density of 32 kg/m3 in tanks supplied with constant aeration.

University of Rhode Island
Graduate School of Oceanography
Kingston, Rhode Island
(Rhodes and Merriner 1973; Rogers 1974)

Studies at the University of Rhode Island were designed to test the use of inexpensive wading pools as alternatives to more permanent concrete or fiberglass tanks. Experiments were conducted in 3 m diameter by 0.6 m deep vinyl lined pools. They were operated as semi-closed systems supplied with recirculated, filtered water and received 95 lpd of make-up water from a well source that was adjusted to 4.7-5.6 ppt salinity. The 3.4-m3 pools were each stocked with 100,000 fry and fed brine shrimp nauplii. When the fingerlings were between 42 and 67 days of age they were fed Tetra-Min flakes. Older fingerlings were fed trout pellets.

The initial objective was to produce yearlings, however the researchers concluded that this system was best suited for fingerling production. They estimated that between 15,000-20,000 fingerlings 20-25 mm in length could be produced in a single pool. The study also showed that vinyl pools were an inexpensive closed system that could be easily setup near the spawning grounds.

Gulf Coast Research Laboratory
Ocean Springs, Mississippi
(Nicholson 1973; McIlwain 1976, 1978)

Probably the most extensive work on the development of intensive culture methods for production of fingerlings has occurred at the Gulf Coast Research Laboratory. Work began in 1969 to stock a North Atlantic Coast strain of striped bass, as part of an effort to establish an anadromous reproducing population of sea-run striped bass along the Gulf Coast.

The culture system developed at this laboratory was comprised of ten banks of five circular fiberglass tanks 2.4 m diameter by 1.2 m deep, with a capacity of 3.9 m3. Four tanks in each module were used for culture and the remaining one was used an up-flow filter.

The circular culture tanks are equipped with a central stand pipe drain and a venturi pipe with openings covered with nytex to retain larvae and fry. Aeration consists of a circular loop around the base of the venturi pipe that produces a bubble curtain of air that helps fry from becoming impinged on the drain screen. Water at 23°C is supplied from a well, pre-treated through pressure sand filters, and used to exchange one volume of the filter tank daily. The up-flow filters have perforated corrugated fiberglass at the bottom that functions as a deflector panel which is covered with 20 cm of clam shell followed by 18 cm of foam rubber. Fry have been stocked in this system at densities ranging from 3-9 fry/liter. The fry are generally fed brine shrimp nauplii from day 5, and brine shrimp nauplii plus wild zooplankton from a 0.1-ha pond after day 10. Commercial trout feeds are supplied by automatic feeders after 15 days. Automatic feeders have also been developed for the brine shrimp and zooplankton (Nicholson pers comm.)

In a study conducted in 1974, 1-day-old fry were stocked at 8.9 fry/l, producing 45.1-mm fingerlings in 60 days with only 0.73% survival (McIlwain 1976). The extremely high mortality rates were caused by _F. columnaris_, which was difficult to control in the filtration pond. A subsequent study conducted in 1976 compared the growth of fry obtained from Virginia with those from South Carolina. When stocked at densities of 3.3 fry/l, the groups from Virginia produced 12-33 mm fingerlings in 27 days, with 38% survival and a growth rate of 0.64 mm/day. The group from South Carolina did not attain this size until after 38 days, showing a slower growth rate of 0.49 mm/day. However, they did have slightly greater survival (50%).

The major cause of mortality in this system was determined to be cannibalism, which became intense after the fingerlings exceeded 20 mm in length. This research also determined that food supply was a major limiting factor in the intensive culture of striped bass fingerlings. To avoid these problems, researchers routinely graded the fish and increased the size and quantity of the pelleted feed. Growth and survival of striped bass in low salinity water was reported to be greater than that of fish cultured in fresh water. In addition, striped bass survival increased as the technicians became more experienced.

Texas Instruments, Inc.
Consolidated Edison Company
Verplanck, New York
(Annual Reports for 1973 to 1976,
Overview 1973-1975)

A 5-year study was conducted between 1973 and 1977 to determine the feasibility of operating a hatchery for the release of striped bass fingerlings into the Hudson River to mitigate power plant entrainment of eggs and larvae. The objective was to refine intensive culture methods for fingerling production and to rear 20,000 7.6- to 12.7-cm fingerlings for mark-recapture studies. A pilot hatchery was constructed for the production of larvae. Some larvae were sent to hatcheries at Durant and Medicine Park, Oklahoma, for production of advanced fingerlings by conventional pond culture methods. Most of the intensive culture experiments were conducted in cooperation with Auburn

University, Southern Illinois University, and the Gulf Coast Research Laboratory.

Results of work in 1973 showed that larval mortality was reduced when they were cultured at a salinity of 2 ppt after 17 days of age rather than in freshwater. Some problems resulting from the fungus Saprolegnia and from parasite infestation have been reduced by use of brackish water in larval culture. Studies in 1974 determined a need to supply adequate quantities of brine shrimp nauplii (300-500 nauplii per larva per day), until the fish became 20-mm fingerlings. They also revealed the importance of feeding nauplii in combination with dry flake food during a 2-week weaning period and the need for grading to reduce cannibalism after fingerlings attain a size of 40 mm.

Initial experiments on intensive culture conducted at the Verplanck Laboratory utilized 3.6 and 4.6 m diameter vinyl-lined pools supplied with quarry water at 13-17°C at a rate of 11.3 lpm adjusted to 2 ppt salinity. Fry reared in aquaria for 28 days to a size of 12-16 mm were stocked in the pools and fed 60 nauplii per larva per day every 3 hours. After 60 days of age this rate was increased to 250 nauplii per fry and supplemented with Tetra-Min dispensed from automatic feeders. After 118 days of culture over 6,000 51-mm fingerlings were produced, however only 6% survived. Some abnormalities were noticed which were attributed to the lack of swimbladder inflation, which caused the fish to swim vertically.

In 1975, studies were conducted in 6.4-m3 stainless steel tanks 3.7 m in diameter. The tanks were equipped with perforated brass cylinder drain collars over the stand pipe and nylon mesh lift-net liners 30 cm deep. Water was introduced at the surface by a spray bar and aeration provided by a bubble collar around the central stand pipe drain. Water flow through the system was 8- to 10-lpm, which produced a peripheral circular velocity of 2 cm/sec. Ten-day-old fry were stocked at densities of 25,000 and 37,000 per tank and fed 60 nauplii per larva per day at 3 hour intervals. This ration was later increased to 300-400 nauplii per larva per day and supplemented with artificial feed. This study produced 7.6- to 12.7-cm fingerlings that were marked with color-coded wire magnetic nose tags and fin-clipped prior to release into the Hudson River.

These studies demonstrated the use of a circular current to aid schooling behavior, facilitate feeding and reduce cannibalism among fingerlings. Also, they showed the importance of a transitional weaning period to dry feeds and the need for grading. The researchers concluded that a dual procedure of culture was preferable, consisting of production of fingerlings in ponds and advanced fingerlings in tanks.

Southern Illinois University
Fisheries Research Laboratory
Gorham, Illinois
(Lewis and Heidinger 1976, 1981)

Cooperative studies with Texas Instruments, Inc. on intensive fingerling production were conducted at Southern Illinois University for 3 years, 1974-1976. Subsequent studies were supported by a sub-contract to UMA Engineers, Inc. from Consolidated Edison Company. In 1974, three rearing systems were tested: 3 m diameter circular pools, 3 m long aluminum raceways, and 1.8 m diameter fiberglass tanks. The study was designed to test various water flow patterns, i.e. circular, linear, downward, and upflow. Larvae were fed nauplii on a 24 hour schedule in subdued light from day 5 through day 30 and maintained at 14-17°C. They were fed 250 nauplii/day until 14 days old and after 20 days old they were fed 500 nauplii/day. Training to dry feeds began from day 12-17 with Tetra-Min and ground salmon starter dispensed from automatic feeders, at which time water temperature was increased to 25°C. By 30 days of age fry were completely weaned onto salmon starter feed, which was fed 12-16 times per day. When less than 40 days old they were fed 10% body weight per day over 10 feedings, and then the feeding rate was reduced to 5% per day for subsequent rearing. Particle size is a critical factor at this stage. Larvae were stocked at 100/l at 1 or 5 days old, to avoid handling shock to the more sensitive 2- to 4-day-old larvae. Fry were shown to have exponential growth for the first 80 days of culture, giving an average growth rate of 2 mm/day. Fry were grown to 51 mm fingerling in 55 days and 120 mm fingerlings in 95 days. Survival was 46% to 5- to 10-cm fingerling size, although nearly half of the fingerlings did not inflate their swimbladder. This phenomenon is commonly observed in intensive culture, but also occurs in pond reared fingerlings at a much lower level (0-55%).

The study of water flow patterns showed a need for an upflow current during the initial 5 days of development, when the larvae have limited swimming ability and have not yet initiated swimbladder inflation for bouyancy control. To meet this need, a unique upflow larval rearing unit was designed. It consisted of a 550-l box constructed of marine plywood. Water was introduced at the bottom of the box and passed through an aluminum plate diffuser positioned 10 cm off the bottom and perforated with 100 3-mm holes. At the top there was an 11 cm high pyramid shaped larval retention screen constructed of 0.5 mm stainless steel screen which was angled 60 degrees inward. The overflow drain had a swivel joint that allowed for adjustment of water level and for draining through the screen or periodic draining above the screen level to remove any surface film. The larval retention screen had 10,000 cm2 surface area, and the water velocity was kept below 0.16 mm/sec to avoid impingment of larvae on the drain screen.

Fry were reared in the upflow system until 10 days old and approximately 6 cm in length. At that time they were transferred to circular production tanks for subsequent growth. It was necessary to transfer larvae before day 12 since the upflow tanks were unfavorable to swimbladder inflation. A bank of 34 insulated, circular tanks was constructed, each of which were 1.8 m in diameter with a volume of 2.0 m3. Each tank was equipped with a central standpipe with bottom drain slots and fitted with a drain screen collar constructed of rectangular mesh metal bars. Both units were operated as closed, recirculating systems with 2-5% make-up water added daily. They were supplied with 14°C well water at a rate of 26 lpm. The recirculation rate was 7.6 lpm during the initial period from 0-5 days old, and then increased to 26.5 lpm thereafter. The circular tanks had a water turn-over rate of four exchanges per day at a flow rate of 11- to 19-lpm. Circular flow was kept at a minimum for the first 4 weeks of culture, and then the angle of the water supply was changed to provide a circular velocity of less than 2 cm/sec. Associated culture system components consisted of a plunge basin aspirator and rotary blower aeration system, a sedimentation basin, an upflow biofilter filled with 1 cm styrene beads and supplied with injected oxygen, pressure sand filters, and ultraviolet sterilization. Flow rate through the filtration system was maintained above 1.4 lpm per kilogram of fish for 1g or smaller fish and was reduced to 0.7 lpm per kilogram for larger fish.

This work has shown the critical need to satisfy the nutritional requirements of fry and fingerlings reared in intensive systems. They demonstrated the need to provide an upflow current during initial development to reduce mortality in 2- to 4-day-old larvae and the subsequent need for a reduced current to accommodate swimbladder inflation. The primary source of mortality in later stages (cannibalism) was limited by periodic grading, and further discouraged by providing a continuous food supply. Other critical factors identified were the deleterious effects of light on feeding, the need for effective aeration and filtration, and the difficulty of isolating the culture system from the biofilters during disease treatments.

Continuing work at the Gulf Coast Research Laboratory and at Southern Illinois University has demonstrated the biological and technical feasibility of intensive culture methods for production of striped bass fingerlings. Other investigators are now attempting to refine these methods.

University of Maryland
Center for Environmental and Estuarine Studies
Horn Point, Maryland
(Krantz 1982 pers. comm)

Investigators at the University of Maryland attempted to utilize large oyster hatchery conical tanks for striped bass fingerling production in the off-season. Sources of supply of seawater and freshwater provided a continuous flow of brackish water to the facility. Water entered the bottom of the cones and was maintained at 0.5–2 ppt salinity until eggs water hardened, which reduced problems with floating eggs and disease. At first feeding, 5-day-old fry were transferred to oyster troughs and salinity increased to 3–5 ppt. After the fingerlings attained a size of 2.5 cm they were converted to full strength seawater, 33 ppt.

Other organizations that are planning to conduct research on the production of striped bass fingerlings in intensive culture systems are:

NMFS Galveston Laboratory, Texas
Moncks Corner Hatchery, South Carolina
Natchitoches Hatchery, Louisiana
Baltimore Gas and Electric Company, Maryland
Consolidated Edison Company, New York
University of Minnesota, Minneapolis

INTENSIVE CULTURE OF ADVANCED FINGERLINGS

Initial efforts to produce fingerlings of 2.5- to 5-cm by intensive culture methods generally resulted in survival rates lower than 10%; survival of striped bass from extensive pond culture averaged about 20%. It was not until the mid 1970s that survival rates for fingerlings reared in intensive culture systems increased to levels exceeding twice (46%) those achieved in ponds (McIlwain 1976; Lewis and Heidinger 1976). Even greater success has been achieved in the culture of advanced fingerlings.

Alabama Department of Conservation
and Natural Resources
Dauphin Island, Alabama
(Swingle 1970, 1972; Powell 1971, 1973)

The focus of this research was to evaluate the

use of cylindrical cages floating in brackish water for the production of advanced fingerlings. The cages were 0.9 m diameter by 1.2 m deep, having a volume of 800-l. They were constructed of wood frames covered with either 0.6 or 1.3 cm mesh vinyl-coated galvanized hardware cloth that was covered with fiberglass window screen. The cages were equipped with styrofoam floats and a feeding ring.

Experiments conducted in 1969 evaluated three stocking densities of 143, 200 and 400 fish per 800-l tank. Fish were stocked at 1.7 g and fed 5% of body weight per day over six feedings. After 2 months of culture, 12-g fingerlings were harvested. Continued studies in 1970 showed that fry could be grown to 15- to 25-cm fingerlings in 5 months on a diet of Purina Trout Chow, with a feed conversion of 3.7-4.5:1 (Swingle 1972). Survival of fingerlings after attaining 10 g was 83.5%. Investigators concluded that floating cage systems were best suited for production of advanced fingerlings from 10 g fingerlings. Problems encountered with this method of culture were fouling of the cage with hydrozoans and mechanical rupture of the cage in rough weather.

Subsequent studies conducted in 1971 and 1972 were designed to determine the appropriate diet and stocking density for fingerlings reared in cages. In experiments on diet and feeding regime, eight 1.3-cm mesh cages were stocked at a density of 30 fry per 800 l. The cages were suspended in brackish water at 19 ppt salinity, and fish were fed Purina Trout Chow or a ground fish plus soybean meal diet at 10% body weight per day during two to four feedings. After 3 months of culture the group fed the trout diet was superior. Fingerlings attained an average weight of 126.1 g, giving a growth rate of 0.98 g/day, survival of 97.4% and a feed conversion of 2.1:1.

A similar experiment conducted in 1972 evaluated three stocking densities of 100, 200, and 300 fry per 800 l. Six 1.3-cm mesh cages were stocked with 136-g fingerlings, two at each of the three densities, and fish were fed Purina Trout Chow at 3% body weight per day. After 60 days of culture at the highest stocking density, the harvest density was 69.6 kg/800 l (0.09kg/l), survival was 99%, and feed conversion was 2.4:1. Fingerlings attained an average size of 226.8 g.

These studies demonstrated that cage culture of striped bass fingerlings in brackish water was feasible. The researchers concluded that the major limiting factor for culture of striped bass in marine environments was bacterial diseases, primarily caused by pseudomonads.

A parallel study on the culture of fingerlings in 2.9-m3 circular fiberglass tanks, equipped with venturi drains and micro-pore diffusers for aeration, gave similarly good results. The tanks were supplied with water at a rate of 76 lpm and the fry were fed Purina Trout Chow. One tank was stocked with 102 9.6-g fingerlings that attained an average weight of 79.9 g in 93 days, with 94.1% survival and a food conversion of 2.0:1. Their growth rate of 0.87 g/day was comparable to that of fish in the most productive cage. The circular raceway design was equally productive as the floating cage. Labor required for feeding and maintenance of fish in the circular raceway was only 25% of that required to maintain fish in the floating cage.

Fish Farming Experimental Station
Bureau of Sport Fisheries and Wildlife
Stuttgart, Arkansas
(Allen 1974)

The success with intensive tank culture methods encouraged additional work on the refinement of these systems. At the Stuttgart Laboratory both circular tanks and raceways were tested. Six circular tanks were constructed of fiberglass with volumes of 1.6 m3. In addition, three rectangular aluminum raceways with volumes of 4.25 m3 were tested. Each tank was supplied with water of 22-24°C at a rate of 30 lpm. Fry were fed floating trout feed with a 40% protein content, five times per day. At the completion of the study 1,136 fingerlings were produced with an average weight of 55.5 g, compared to 11.6 g for fry reared in ponds.

In a parallel experiment, 118 fry with an average weight of 11.7 g were stocked in raceways receiving water of 29°C at a rate of 60 lpm. They were fed trout feed and floating catfish feed having a 30% protein content. At the end of the summer growing season this system produced 86 advanced fingerlings (0.7% survival) with an average weight of 178 g and a feed conversion ratio of 2.2:1. The projected growth rate of yearlings suggest that they would have exceeded 500 g by the end of the second summer.

Central Valley's Hatchery
California Department of Fish and Game
Elk Grove, California
(Cochran pers. comm.)

The production of striped bass fingerlings was initiated at one of California's warmwater fish hatcheries in 1972. This facility was originally constructed in 1938 for the culture of smallmouth bass. Therefore, it was equipped with fingerling production ponds to which troughs and concrete raceways were added in 1973. After 4-8 weeks of pond culture, the resulting 23-mm (0.2 g) fingerlings were stocked at a density of 6,000 per 300-l trough. The troughs were supplied with well water passed through a reservoir pond where it was aerated and heated to about 24°C. The fingerlings were fed moist and dry salmon diets and graded frequently to reduce cannibalism. Fingerlings were grown in the troughs for 10-14 days to an average size of 0.8 g after which the density is reduced to 3,000 fingerlings per trough. Fingerlings were transferred in July from the rearing troughs to 12.9-m3 outdoor concrete raceways and stocked at a density of 6 kg/m3 for final grow-out to yearlings. The raceways were supplied with well water at 20°C which was aerated by an aspirator. The fingerlings were fed commercial salmon feed at 3% body weight over three feedings per day. Yearlings were harvested in April at an average weight of 252 g per fish. Advanced fingerling mortalities in this system were usually minimal. The approximately

150,000 yearlings produced each year were used for stocking several freshwater reservoirs in California and coastal areas of Southern California. Plans to increase the enhancement efforts in San Francisco Bay and the Sacramento-San Joaquin Delta will require that annual production at this hatchery be doubled over the next 3 years.

Southeastern Fish Cultural Laboratory
U.S. Fish and Wildlife Service
Marion, Alabama
(Parker 1979a, 1979b, pers. comm.)

Considerable research has been conducted at the USFW Laboratory at Marion, Alabama, on high density pond culture of channel catfish in ponds equipped with air-lift pumps. Small ponds of 0.02 to 0.01 ha were found to be easily managed and could support standing crop of approximately 10,000 kg/ha of channel catfish. Preliminary studies suggest that the maximum standing crop for striped bass may be significantly increased in ponds with airlift pumps. Current studies in cooperation with the Edenton and Natchitoches Hatcheries indicate that both the standing crop (62-77 kg/ha) and the survival rate (25-66%) of hybrid fingerlings reared in aerated ponds will more than double levels achieved by conventional pond culture methods (26-49 kg/ha and 10-46% survival).

Another area of research at this facility concerns the use of high-density silo culture techniques for production of fingerlings or food fish and for holding domestic broodstock. Two subsurface semi-closed silos were installed in a water reuse system in 1978. The silos were 3 m dia. by 6 m deep with concave bottoms and a capacity of 42.0 m3. Aeration and current patterns were maintained by internal airlift pumps. The recirculation rate was 2,100 lpm, which resulted in a turnover of the water every 20 minutes, or three times per hour. The waste water was removed from the silo through a 15 cm diameter drain in the bottom, which was connected to an external airlift riser. Make-up water was added at the top of the silo at a rate of 3 lpm, which was approximately a 10% exchange per day. Waste water from the silo passed through two primary clarifiers, designed as incline plate separators, at a rate of 26 silo volumes per day (750 lpm). These units were triangular shaped boxes having 62 m2 of surface area on a series of corrugated fiberglass plates positioned at a 45 or 60 degree angle. Water flow was downward at a rate of 112 cm/min, which facilitates removal of particles with a settling rate greater than 3 cm/min. The average settling rate of particulate waste from striped bass was found to be 16.7 cm/min, and the plate separators removed about 90% of the settleable solids.

From the primary clarifiers the water entered a four stage rotating biological contactor (RBC) at a rate of 712 lpm. The remainder of the water was returned to the culture silos. The RBC contained 1,400 m2 of surface area on 2 m diameter discs that rotated at 2 rpm. The hydraulic load for this unit was initially 411 liters/m2 but has been increased to 733 liters/m2 of filter substrate. From the RBC, water passed through a secondary clarifier to remove particles with settling rates greater than 0.8 cm/min and was then returned to the silos.

Striped bass, striped bass hybrids and channel catfish have been grown in this system. Dissolved oxygen became limiting when the load per silo exceeded 1360 kg of channel catfish. Injections of liquid oxygen will be required for fish densities greater than 32 kg/m3 (2lb/ft3).

Subsurface silos have the benefits of providing thermal stabilization and low head differences between the culture tank and the filtration components. Unfortunately, the latter feature also makes it difficult to drain and harvest fish from the silos. Harvesting has been accomplished with a lift screen from the bottom, by chemical treatment and a 3-m diameter by 3-m deep net. The fish cannot be graded easily and considerable cannibalism has been observed with fingerling fish.

One silo in this system has been covered with a dome and used to over-winter _Tilapia sp._ The silos have been equipped with automatic feeders and an electrical grid system has been used to replace the mechanical screen at the bottom of the tank.

Gallatin Steam Plant
Tennessee Valley Authority
Gallatin, Tennessee
(Collins et al. 1982; Schweinforth pers. comm.)

Studies by TVA at the Gallatin facility have focused on the use of thermal effluent for high-density culture of channel catfish in concrete raceways. In 1981 a preliminary experiment was conducted on the potential for hybrid striped bass culture in this system. A 4.4-m3 section of one of the 17.4-m3 raceways was stocked with 385 hybrids averaging 3.1 g and 5 cm in length. Thermal effluent was supplied at a rate of 387 lpm which resulted in a turnover every 16 minutes. The water averaged 26.7°C and was aerated by the injection of liquid oxygen. The fingerlings were fed a 36% protein floating catfish feed and a 45% protein sinking trout feed over four daily feedings. After 117 days, 336 advanced fingerlings were harvested with an average weight of 127 g, length of 20.8 cm, growth rate of 0.95 g/day, survival of 87.3%, food conversion of 1.94:1, and harvest density of 9 kg/m3.

Results from this study indicated that hybrid striped bass preferred the 45% protein sinking pellet when compared to the 30% protein floating pellet. Some problems due to crowding and _F. columnaris_ were observed. This study showed that the growth rate of hybrid striped bass cultured in thermal effluent was accelerated, yet, feed conversion remained favorable.

INTENSIVE CULTURE OF FOOD FISH

Marine Protein Corporation
Homestead, Florida
(Stevens 1979 pers. comm.)

Extensive culture of striped bass to a marketable size of 0.5-1.0 kg in intensive systems was first attempted in 1973 in a commercial pilot production facility built by Marine Protein Corporation. Fry were cultured in 208-l drums and 1.2-1.9-m3 horizontal raceways. Over 300,000 fingerlings were produced from these systems. Fingerlings and market size fish were grown in six silos of varying heights and volumes, 70.4-136.6 m3. The flow rate was adjusted to provide 5.7 lpm for each kg of fish in the system. The silos were stocked with 244,000 fingerlings of 5-17.8 cm. Of the remaining fingerlings, 60,000 were sold for stocking and 4,536 kg were grown to advanced fingerlings size (25.4-28 cm) under contract with Texas Instruments, Inc. for mark-recapture studies in the Hudson River.

Of the fingerlings stocked in the six silos, 9,072 kg of 36- to 46-cm food fish were produced. Problems encountered included mortalities caused by cannibalism, disease, and pesticide contamination of the water supply. These systems were both labor and capital intensive, and required considerable amounts of energy. After 3 years of investigation these systems proved to be uneconomical. However, with the improved technology available today and more favorable market conditions, this method of culture may now be feasible.

New York Ocean Science Laboratory
Montauk, Long Island, New York
(Valenti 1976)

At the NYOSL in 1974, floating cage culture systems were evaluated in the colder waters of the Northeast. The cages were cubical, having a volume of 1.82 m3. They were constructed of 16 gauge vinyl-coated galvanized wire with a 2.5 x 1.3 cm mesh. The initial study was conducted to evaluate stocking densities. The fingerlings were fed Central Soya Trout feed at 8% body weight per day. Fingerlings stocked in June had 94% survival rate until February, when unseasonally cold temperatures killed 75% of the fish. Surviving fingerlings were overwintered in 200-l indoor pools at 16°C. A second group stocked in June at densities of 28-50/m3 grew to 383-420 g, with feed conversions of

1.4-1.8:1. The rate of growth was shown to be a function of stocking density. At harvest the standing crop was 19 kg/m3. Investigators concluded that 0.25-0.5 kg marketable size fish could be grown in one year.

Floating cages were demonstrated to require less space and power than did intensive systems previously tested. However, new problems were introduced related to the high maintenance of cages that frequently became fouled with marine organisms, and also the difficulty of harvesting with a hoist from a floating platform.

Center for Marine Studies
San Diego State University
San Diego, California
(Van Olst and Carlberg 1978; Van Olst et al. 1979)

In 1970 Sea Grant began to support research at SDSU on the growth rate of crustaceans and finfish cultured at elevated temperatures in the thermal effluents from coastal power plants. Preliminary work on striped bass culture was conducted in 1976 at the Encina Power Plant of San Diego Gas & Electric Company, under funding from the Southern California Edison Company. These studies were performed in conjunction with a program of the California Department of Fish and Game to release yearling striped bass in coastal estuaries in hopes of developing an anadromous population of stripers in Southern California. Advanced fingerlings of about 30 g were cultured in ambient temperature and thermal effluent seawater. Results showed that striped bass could be acclimated to full strength seawater and grown at elevated temperatures.

A second study was conducted in 1977 at the SDSU Aquaculture Laboratory located at the Scripps Institution of Oceanography. Advanced fingerlings with an average weight of 91.5 g were stocked at a rate of 4.2 kg/1 into three 1.1-m3 circular fiberglass tanks. Ambient temperature seawater with a mean of 16°C and a range of 12-21°C was supplied to each tank at a rate of 12 lpm. In one year the fingerlings were grown to a weight of 605 g and a length of 35 cm. Survival was 100%, with an average growth rate of 1.7 g/day, a harvest density of 27.5 kg/1, and a feed conversion of about 2:1. The relatively low standing crop at harvest resulted from the relatively low initial stocking density.

Based on the favorable results of these two preliminary studies, an experiment was conducted in 1978 to evaluate growth and survival at different temperatures and feeding levels. Advanced fingerlings with an average weight of 38.2 g were stocked in ten 760-1 rectangular fiberglass raceways at densities of 2.5 g/1. The raceways were equipped with a center divider and air-lifts along the divider and around the perimeter to provide aeration and a circular current. Four raceways were supplied with mixed thermal effluent and ambient temperature seawater at 24°C, four at 22°C and two at ambient 18°C. Fingerlings in two of the four raceways at 22°C were fed 1.5% body weight per day, all others received 3% body weight per day of Purina floating Trout Chow. After 4 months the highest specific growth rate (1.6% per day) and feed conversion ratio (2.2:1) were recorded for fish cultured at the highest temperature, 24°C. For fish reared at 18°C growth rates was 0.98% per day and feed conversion was 2.6:1 whereas fish reared 22°C grew at 1.3% per day and had a feed conversion of 2.4:1. The effect of lower feeding levels (1.5% per day) resulted in a slower growth rate, 1.0% per day.

A related study conducted in 1979 involved the evaluation of floating cages in Southern California bays for the culture of advanced fingerlings prior to release for enhancement of the marine recreational fishery. Wood frame walkways supported by styrofoam floats were used to suspend 20-m3 mesh knotless nylon cages. Approximately 1,000 70-g fingerlings were stocked into cages in San Diego and Mission Bays. Previous studies demonstrated that 70-g fingerlings could be grown to 1.0 kg in less than a year. The need to vaccinate fingerlings prior to culture in full-strength seawater was demonstrated by the considerable mortalities attributed to *Vibrio* infection.

Mercer Generating Station
Public Service Electric and Gas Company
Trenton, New Jersey
(Guerra and Godfriaux 1978; Eble and Godfriaux pers. comm.)

A similar study on the use of a freshwater thermal effluent from a power plant was conducted in 1978 by researchers at Trenton State College under contract from the PSE&G Company. Three rearing trials were attempted. The first attempts were in a vinyl lined earthen raceway and a 246-m3 pond. The raceway was stocked with 4,000 advanced fingerlings averaging 20 cm in length, at a density of 88 fish/m3. The raceway was supplied with thermal effluent at a rate that produced one turn over every hour. The pond was stocked with 6,000 fingerlings of the same size. Fingerlings in both systems were fed trout pellets.

The objective of this study was to identify a species that could be grown at this facility during the warm summer season to augment their culture of rainbow trout during the cool season. Due to the simultaneous shut-down of both condenser units of the power plant, culture temperatures dropped to ambient temperatures of the Delaware River. This sudden thermal shift resulted in an infestation of the fungus *Saprolegnia* and less than 3% of the fingerling striped bass survived.

The third study was conducted using 1.5-m3 square fiberglass tanks with rounded corners. The water temperature was 25.6°C and the flow rate was 20 lpm. Trout pellets were fed at a rate of 10% body weight per day. Considerable mortality was caused by fungal and protozoan diseases, as well as by cannibalism. The investigators concluded that striped bass culture at this location was unfeasibile due to the susceptiblity of striped bass to pathogens commonly found in the Delaware River.

Multi-Aquaculture Systems
Amagansett, New York
(Valenti pers. comm.)

Success of the prior study at the Ocean Science Laboratory stimulated the interest of developing a commercial striped bass cage culture operation on Long Island. In the late 1970s MAS constructed twenty-two 45 4-m3 vinyl-lined circular pools. A seawater well supplies water at 10°C and 26 ppt salinity at a rate of 117 lpm to each pool. In addition, several floating 41-m3 cages were constructed. The cages were stocked with 60 day old fingerlings at a density of 100,000 per cage and fed ground trash fish three times per day. Bay water had a salinity of about 28 ppt and a temperature of 20°C in the summer. Fingerlings grew from 0.5 g to about 75 grams from late April to November. Each cage produced approximately 907 kg of advanced fingerlings with an 80% survival. The fingerlings were then stocked for overwintering in the pools at a density of 5,000 fish per pool. Little growth occured at the low winter temperatures. The following spring the fingerlings were restocked into the cages. In December, 907 kg of 220- to 340-g food fish were harvested from each cage and these 18- to 20-month-old striped bass were marketed at a price of $4.00 per pound.

The problems encountered with this production method have included fouling of the nets in the summer (nets had to be brushed every few days) tearing of the nets by predators, and slow growth of the fish due to low winter temperatures.

DOMSEA Farms Inc.
Marine Resources Research Institute
Charleston, South Carolina
(Williams 1981; Williams, Lindberg,
and Sandifer pers. comm.)

Similar cage culture experiments were conducted in 1978 and 1979 at the Stono River Marina in South Carolina. Two 13.6-m3 cages were constructed from wood frames supported by styrofoam floats from which either 0.6 or 1.9 cm mesh knotless nylon nets were suspended. Temperature in the bay averaged 28°C but varied between 7-31°C, and salinity had a mean of 22 ppt with a range of 11-28 ppt. Fingerlings were fed commercial salmon feed and ground trash fish at 5% body weight once per day.

The initial study resulted in over 78% mortality in the first 40 days of culture, due to the inability of the fingerlings to successfully make the transition from fresh fish to pelleted feed. Problems also were encountered with *Vibrio* disease and cannibalism. Hybrids appeared to be less affected than striped bass. In the second attempt hybrids were vaccinated for *Vibrio* and fed a combined diet of fish and pellets. Two cages were stocked with 1,070 40-day-old fingerlings weighing 1.7 g each. Survival was 74.6% when the fish were 380 days old and they averaged 523 g. The growth rate was 1.37 g/day, with a standing crop of 12 kg/m3, and a feed conversion of 3.5:1.

A third experiment was conducted in 1979 in which three cages were stocked with 900 3 g fingerlings at a density of 66 fish/m3. Survival was 88% when the fish were 380 days old and they averaged 310 g. The growth rate was 0.81 g/day, a standing crop of 16 kg/m3, and a feed conversion of 2.1:1. The slower growth rate and lower feed conversion was attributed to limited food at the lower feeding level of 1-2%/day in this experiment.

Problems encountered were again related to biofouling, requiring the net to be changed every 4 weeks, and damage to nets caused by blue crabs.

This study showed that a pan-size striped bass could be produced from April to December, and marketed prior to the onset of cold winter water temperatures.

Aurora Field Station
North Carolina State University
Aurora, North Carolina
(Woods 1981; Kerby 1982 pers. comm.)

In 1980 preliminary studies were conducted to evaluate high-density pond culture. Fingerlings stocked at 20 g were grown to 351 g in 13 months, with a survival of 93%, and a standing crop of 4,886 kg/ha. Problems encountered were caused by handling stress and bird predation.

Studies in 1981 evaluated pools and cages for use in striped bass culture. Four 1.6-m3 circular tanks were constructed from 122-cm corrugated galvanized metal used for grain silos, and were fitted with vinyl pool liners. The pools were equipped with a central stand pipe drain and airlifts for circulation. Well water was supplied at a rate of 60 lpm. Fingerlings were fed 40% protein trout feed. The pools were stocked with 53-g hybrids at densities of 0.9, 3.2, and 4.5 kg/m3. At harvest fish in the high density tanks averaged 275 g with a standing crop of 3.9 kg/m3, and 74% survival.

A parallel study conducted in nine cages stocked at 100, 150, or 200 fish/m3 gave similar results. Upon harvest the high stocking density cages produced 310 g fish at a standing crop of 59.7 kg/m3, and 74% survival.

Problems encountered in this study were *F. columnaris* disease, cannibalism, low dissolved oxygen levels, high maintenance requirements, and lack of control over temperature flucuations. Nevertheless, this study did demonstrate that a marketable size hybrid striped bass could be produced in 15-18 months when cultured at high density.

Aquatic Systems Incorporated
San Diego, California
(Van Olst et al. 1980, 1981; Carlberg et al. 1981)

The principals of ASI, while at SDSU, directed the research previously described on the use of thermal effluent in the intensive culture of striped bass (Van Olst and Carlberg 1978, 1979). Based on the success of these studies they began a program in 1980 to develop a commercial striped bass culture operation.

Although the results from the earlier grow-out experiments were encouraging, these studies depended upon obtaining fingerlings from the state hatchery. Since there was no commercial source of fingerlings, the first step was to begin development of an independent supply of larvae and fingerlings. Based on previous success with the culture of crustacean and finfish larvae under controlled environmental conditions, efforts were focused on the modification of these systems for rearing striped bass. The development of intensive fingerling culture techniques was a cooperative effort between ASI, the University of California at Davis, and the Department of Fish and Game.

The first study involved the use of fiberglass cylindrical tanks with conical bottoms and volumes of 375 l. The units were designed as recirculated closed systems to facilitate environmental control. Water temperature was maintained at 16-19°C, salinity at 10-12 ppt, and ammonia levels controlled by filtration and water exchange. The tanks had a common central standpipe drain and water circulation apparatus that provided an upflow current. A drain collar covered with Nytex screen retained the larvae and their food supply while an aeration ring produced a bubble curtain that kept the screen from clogging. Larvae were fed nauplii and moist salmon diet. Initial results were successful, with survival exceeding 90% from hatch to 10-day-old fry. Observed mortalities were caused by cannibalism and starvation resulting from under feeding. Upflow circulation allowed for control over the distribution of larvae and nauplii, and facilitated cleaning and maintenance. This culture unit was demonstrated to be useful for hatching eggs, rearing larvae to initial feeding, and for the weaning of fry to artificial diets with a minimum of labor.

Refinement of the design and operation of the tank was continued in 1981. New tanks were fabricated that had an improved shape, color, and size, but incorporated the water upflow and screen design of the earlier prototype. These tanks, which are still in use by ASI, are an ogive shape with a parabolic bottom. They also have a white bottom to assist observation from above by the culturist. The ogive unit has a volume of 100 l and can hold 100,000 eggs for hatching and development prior to feeding. Fry were fed hatched, de-capsulated, brine shrimp nauplii to reduce the time consuming effort required for separating cyst shells from hatched nauplii, and to eliminate problems associated with ingestion of unhatched cysts. Nauplii were moved to the ogive tank from a holding reservoir by a metering pump at 15 minute intervals 24 hours a day. Later stages of fry were continuously fed moist salmon diets by use of automatic feeders. Work in 1981 showed that a flow rate of 7.6 lpm provided sufficient flushing, that acclimation to a salinity of 10 ppt, 12 hours after hatch, significantly reduced mortality, and that gentle aeration, dim light, and delayed feeding may contribute to increased survival and normal inflation of the swinbladder.

A prototype modular spawning system was fabricated and used in field spawning operations in California, Oregon, and Louisiana. From this program, three separate strains of striped bass have been maintained and grown to 2-8 kg for use as domestic broodstock. Related studies on out-of-season spawning are in progress with these and other captive adults.

The evaluation of using thermal effluent seawater for striped bass culture was continued in 1981. A cooperative study with SDSU and San Diego Gas & Electric was designed to refine methods for use in future commercial mariculture operations. The work was conducted at the Encina Power Plant of SDG&E. Experiments were conducted on the effects of temperature, salinity, stocking density, feeding level, and vaccination on the growth and survival of striped bass. Approximately 6,000 advanced fingerlings weighing an average of 40 g were used in these studies.

Results from these experiments showed that specific growth rates of 4.58%/day were achievable at temperatures of 24-28°C and reduced to 2.98%/day at 20°C. Salanities of 0, 20, and 33 ppt had no significant effect on growth or survival. Neither temperature nor salinity significantly affected feed conversion.

In the *Vibrio* immunization experiment, 20-g fingerlings exhibited antibody agglutination reactions within 3 weeks of vaccination. Larger 30-g fingerlings showed an immune response by the eighth week.

The study on stocking density compared four

levels: 8, 16, 24, and 32 kg/m3. Results showed that growth and feed conversion were reduced at the highest density tested. However, yield was also greatest at the highest stocking density.

Studies at ASI's laboratory in San Diego and at its pond culture facility in Central California have been conducted under commercial operation conditions. Research on intensive culture systems was conducted in a series of 946-1 circular fiberglass tanks, operated as semi-closed systems with a recirculation rate of 19 lpm, and a turn-over rate of about half the tank volume daily. A majority of the tanks were 2.7-4.2 m3 in volume and operated as closed systems. The systems were equipped with inclined plate separators, high rate sand filters, and bio-trickle filters. With these filtration components, culture densities of 40 kg/m3 were routinely maintained. Most fingerlings and larger fish were fed commercial trout feeds. A study that compared floating to sinking pellets showed no significant differences on growth between groups. Other experimental feeds are currently being tested on several life stages. Some results have shown that although maximum growth was near 26°C, optimum temperature for both growth and feed conversion was closer to 24°C. Preliminary work with striped bass x white bass hybrids has shown that under optimum conditions hybrid fingerlings grew at a rate of 5.3%/day, which was nearly twice that of striped bass. Hybrids maintained feed conversion efficiencies (1.3:1) equal to or better that found for striped bass. Hybrids also had an increased resistance to stress induced diseases. Generally it has been found that a one-pound marketable hybrid can be grown in 9-12 months, compared to 12-15 months for striped bass.

Mortality for striped bass and hybrid fingerlings larger than 2.5 cm is generally less than 2% per year, except during periods involving handling or mechanical failures. Cannibalism is a problem with fry, but not with fingerlings if they are well fed. Most significant disease-related mortalities have been the result of infections by _Aeromonas_,

Pseudomonas and *F. columnaris* bacteria. Fungus infestations and bacterial attacks of *Vibrio* have been controlled with KMnO4 and vaccinations, respectively.

Future plans of ASI involve the development of intensive commercial production facilities at sites where thermal effluent or geothermal warmwater is available. These systems will consist of a series of large circular tanks, equipped with oxygen injection and sophisticated filtration for removal of suspended solids and ammonia.

CONCLUSIONS

Intensive culture of striped bass on a commercial scale requires considerable capital for construction and operation at an economical level. Major constraints to implementation of the developed technology are the lack of suitable sites and difficulty in obtaining financing.

The future development of intensive culture systems for striped bass will depend upon solving three major limiting factors. First, additional research is needed on the development of domestic broodstock to facilitate a year-round supply of fingerlings of a strain that is well adjusted to culture conditions. Second, increased production of fingerlings by intensive methods will require improved diet formulation, and better control of cannibalism and swimbladder inflation. Third, techniques and procedures must be developed to reduce handling stress and to control infectious diseases.

The biological and technological feasibility of intensive striped bass culture has been proven. What is now needed is to demonstrate that the operation of a commercial-scale facility will be economically viable. There is good evidence that a striped bass culture industry will soon be developed.

REFERENCES

Anon. 1974-1978. Feasibility of Culturing and Stocking Hudson River Striped Bass. Ann. Reports for 1973-1977. Texas Instruments Inc. Science Services Division, Dallas, Texas.

Anon. 1977. Feasibility of Culturing and Stocking Hudson River Striped Bass. An Overview, 1973-1975. Texas Instruments Inc. Science Services Division, Dallas, Texas. 81p.

Allen, K.O. 1974. Notes on the Culture of Striped Bass in Tanks and Small Raceways. Prog.Fish Cult. 36(1):60-61.

Anderson, J.C. 1966. Production of Striped Bass Fingerlings. Prog. Fish Cult. 28(3):162-164.

Bailey, W.M. 1974. An Evaluation of Striped Bass Introductions in the Southern United States. Proc. Ann. Conf. S.E. Assoc. Fish and game Comm. 28:54-68.

Bayless, J.D. 1972. Artificial Propagation and Hybridization of Striped Bass, *Morone saxatilis* (Walbaum). South Carolina Wildlife and Marine Resources Department. 135p.

Bonn, E.W. (ed.). 1976. Guidelines for Striped Bass Culture. Striped Bass Culture. Striped Bass Committee, So. Div., Amer. Fish Soc. 103p.

Braid, M.R. and E.W. Shell. 1981. Incidence of Cannibalism among Striped Bass Fry in an Intensive Culture System. Prog. Fish Cult. 43(4):210-212.

Braschler, E.W. 1974. Development of Pond Culture Techniques for Striped Bass *Morone saxatilis* (Walbaum). Proc. Ann. Conf. S.E. Assoc. Fish and Game Comm. 28:44-48.

Carlberg, J.M., J.C. Van Olst, M.J. Massingill, T.A. Hovanec, M.D. Cochran, and S.I. Doroshev. 1981. Intensive Culture of Striped Bass Larvae at Central Valleys Hatchery. Progress Report to the California Department of Fish and Game. 15p

Collins, C.M., C.L. Burtom, and R.L. Schweinforth. 1981. Raceway Culture of Striped Bass x White Bass Fingerlings Utilizing Power Plant Waste Heat. Fourth ann. Proc. Catfish Farmers of America. pp.39-40.

Colura, R.L., B.T. Hysmith and R.F. Stevens. 1976. Fingerling Production of Striped Bass, *Morone saxatilis*, Spotted Seatrout, *Cynoscion nebulosus*, and Red Drum, *Sciaenops ocellatus* in Saltwater Ponds. Proc. World Mariculture Society 7:79-92.

Guerra, C.R. and B.L. Godfriaux. 1978. Power Plant Waste Heat Utilization in Aquaculture. Second Annual Report. Public Service Electric and Gas Company. pp.23-24 and 76-90.

Inslee, T.D. 1977. Holding Striped Bass Larvae in Cages Until Swim-up. Proc. Ann. Conf. S.E. Assoc. Fish and Wildlife Agencies 31:422-424.

Kelley, J.R. Jr. 1967. Preliminary Report on Methods for Rearing Striped Bass, *Roccus saxatilis* (Walbaum), Fingerlings. Proc. Ann. Conf. S.E. Assoc. Game and Fish Comm. 20:341-356.

Kerby, J.H. and E.B. Joseph. 1978. Growth and Survival of Striped Bass and Striped Bass x White Perch Hybrids. Poc. Ann. Conf. S.E. Assoc. Fish and Wildlife Agencies 32:715-726.

Kerby, J.H., L.C. Woods, III, and M.T. Huish. 1982. A Review of Methods, Advances and Problems Associated with Culture of the Striped Bass and Its Hybrids. 64p. (in prep).

Krantz, G.E. 1982. Current Mid-Atlantic Fish Culture Operations. Striped Bass Culture at the University of Maryland. Proc. 4th Ann. Maryland Aquaculture Conf. (unpublished).

Lewis, W.M. and R.C. Heidinger. 1976. Striped Bass Rearing Experiments 1976. Southern Illinois University. 197p.

Lewis, W.M. and R.C. Heidinger. 1981. Tank Culture of Striped Bass Production Manual. Illinois Striped Bass Project. I.D.C. F-26-R. Southern Illinois University. 115p.

McIlwain, T.D. 1976. Closed Circulating System for Striped Bass Production. Proc. World Mariculture Society 7:523-534.

McIlwain, T.D. 1978. Rearing and Stocking Striped Bass - Mississippi Gulf Coast. Ann. Prog. Report on Project AFCS-6-18/2. Gulf Coast Research Laboratory. 102p.

McIlwain, T.D. 1980. Striped Bass Restoration Program - Mississippi Gulf Coast. Ann. Prog. Report on Project AFCS-7-1. Gulf Coast Research Laboratory. 88p.

Nicholson, L.C. 1973. Culture of Striped Bass (*Morone saxatilis*) in Raceways under Controlled Conditions. Proc. Ann. Conf. S.W. Assoc. of State Game and Fish Comm. 1973.

Parker, N.C. 1979a. Striped Bass Culture in Continuously Aerated Ponds. Proc. Ann. Conf. S.E. Assoc. Fish and Wildlife Agencies 33:353-360.

Parker, N.C. 1979b. An Air-operated Fish Culture System with Water-Reuse and Subsurface Silos. Bio.-Eng. Sym. Fish. Cult. 13p.

Powell, M.R. 1971. Striped Bass, *Morone saxatilis*,

Production to Establish Commercial Stocks in Alabama Estuaries. Ann. Prog. Report on Project AFC-3-2. Alabama Department of Conservation & Natural Resources. 26p.

Powell, M.R. 1973. Cage and Raceway Culture of Striped Bass in Brackish Water in Alabama. Proc. Ann. Conf. S.E. Assoc. Game and Fish Comm. 26:345-356.

Rhodes, W. and J.V. Merriner. 1973. A Preliminary Report on Closed System Rearing of Striped Bass Sac Fry to Fingerling Size. Prog. Fish Cult. 35(4):199-201.

Rogers, B.A. 1974. Growing Striped Bass in Captivity. Maritimes 18(4):1-3.

Shell, E.W. 1972. Effects of Increased Water Hardness, Source of Fry, and Age at Stocking on Survival and Growth of Striped Bass Fry in Ponds. Ann. Prog. Report for Project AFC-4, Segment 2. Auburn University. 31p.

Snow, J.R. 1977. Feasibility of Increasing the Yield of Striped Bass Fingerlings Through Use of Intensive Culture Methods. Ann. Prog. Report for Project AFC-9-1, Segment 1. Auburn University. 27p.

Snow, J.R. 1979. Feasibility of Increasing The Yield of Striped Bass Fingerlings Through Use of Intensive Culture Methods. Ann. Prog. Report for Project AFC-9-2, Segment 2. Auburn University. 46p.

Stevens, R.F. 1966. Hormone-Induced Spawning of Striped Bass for Reservoir Stocking. Prog. Fish Cult. 218(1):19-28.

Stevens, R.F. 1979. Striped Bass Culture in the United States. The Comm. Fish Farmer & Aquaculture News. 5(3):10-14.

Swingle, W.E. 1972. Alabamas Marine Cage Culture Studies. Proc. World Mariculture Society 3:75-81.

Swingle, W.E. 1970. Experiments in the Culture of Marine Species in Floating Baskets. Ann. Prog. Report on Project 2-86-R-1. Alabama Department of Conservation. 26p.

Tatum, B.L., J.D. Bayless, E.G. McCoy, and W.B. Smith. 1965. Preliminary Experiments in the Artificial Propagation of Striped Bass, *Roccus saxatilis*. Proc. Ann. Conf. S.E. Assoc. Game and Fish Comm. 19:374-389.

Valenti, R.J., J. Aldred, and J. Liebell. 1976. Experimental Marine Cage Culture of Striped Bass in Northern Waters. Proc. World Mariculture Society 7:99-108.

Van Olst, J.C. and J.M. Carlberg. 1978. Beneficial Use of Thermal Effluent in Aquaculture. Final Report for Phase III. Research Contract No. U 2585907. Research and Development Department, Southern California Edison Company. pp.494-518.

Van Olst, J.C., R.M. Ford, J.M. Carlberg, and W.R. Schmitt. 1979. Re-establishment of Anadromous Fishes in Southern California. Ann. Report for Project No. R/F-42. Univ. of Calif. Sea Grant College Prog. 3p.

Van Olst, J.C., J.M. Carlberg, M.J. Massingill, T.A. Hovanec, M.D. Cochran, and S.I. Doroshev. 1980. Methods for Intensive Culture of Striped Bass Larvae at Central Valleys Hatchery. Prog. Report to the Calif. Dept. of Fish and Game. 17p.

Van Olst, J.C., J.M. Carlberg, M.J. Massingill, T.A. Hovanec, A. Fusfeld, and S. Nakhoul. 1982. Preliminary Investigations on the Culture of Striped Bass in Seawater. Proc. World Mariculture Society 13:14p. (in press).

Williams, J.E., P.A. Sandifer, and J.M. Lindbergh. 1981. Net-Pen Culture of Striped Bass x White Bass Hybrids in Estuarine Waters of South Carolina: A Pilot Study. Proc. World Mariculture Society 12(2):98-110.

Wirtanen, L.J. and R.H. Ray. 1971. 1970 Report on the Development of Essential Requirements for Production. U.S. Fish and Wildlife Service. 37p.

Woods, L.C., III, J.H. Kerby, M.T. Huish, and G.M. Gafford. 1981. Circular Tank for Intensive Culture of Hybrid Striped Bass. Prog. Fish Cult. 43(4):199-200.

High Density Culture of White Bass x Striped Bass Fingerlings in Raceways Using Power Plant Heated Effluent

Charles M. Collins, Gary L. Burton
and Ross L. Schweinforth

Culture of the striped bass X white bass hybrids and the reciprocal cross, is currently generating a great deal of interest in various state game and fish agencies, federal organizations and private enterprises due to the success of hybridization, the hybrids value as a sport and table fish, its faster growth rate than the parent species for the first 18 months, and its adaptability to intensive culture (Stevens 1964, 1983; Bishop 1967; Logan 1967; Collins, Burton and Schweinforth 1982b; Woods et al. 1983).

The culturing of a relatively small number of hybrids at Gallatin during 1981 and the interest shown by private industry stimulated further culture investigations.

METHODS

The Gallatin Waste Heat Aquaculture Facility is located on the bank of TVA's Gallatin Steam Plant discharge canal near Gallatin, Tennessee. The power plant uses once-through cooling with an average Δ T of approximately 12°F.

Five 3 x 8 x 1 ft wire mesh floating cages suspended in three 614 ft^{3} (50 x 4 x 4 ft outside dimensions) concrete raceways were stocked on May 19, 1982 with 21,560 hybrids weighing 1,691 fish/lb. A water flow rate of approximately 100 gpm (complete exchange every 45 minutes) was maintained during cage culture. Sub-samples from two of the 14 oxygenated plastic bags were taken prior to stocking to determine the number of fish stocked. Fish were kept in cages for 20 days. A 38 percent protein sinking trout ration was pulverized and hand fed *ad libitum* four times daily at the beginning of the culture period. Feeding frequency was increased to nine times daily after it was apparent that the fish would feed more often. Trout ration was replaced with a 50 percent protein trout starter ration (size 2) because of the inconvenience of preparing the 38 percent protein ration in a blender, gradual decrease in feeding activity, daily mortalities and cannibalism.

Four 614 ft^3 concrete raceways were stocked from cages on June 7 through June 12 with 9,210 hybrids. Three raceways were stocked at 2.6 fish/ft^3 (1,535 fish/raceway) and the remaining raceway at 5.1 fish/ft^3 (3,070 fish) (Table 1). Each raceway was divided into two separate culture units (307 ft^3) by a wire mesh divider. A flow rate of 300 gpm (complete exchange every 16 minutes) was maintained in each of the raceways. Fish averaged 361/lb at stocking. A 40 percent protein sinking trout ration (size 4) was hand fed *ad libitum* four times daily initially and gradually decreased to twice daily after the hybrids approached four inches in total length. On July 5 a 37.5 percent protein floating trout ration was fed with the sinking ration. This feeding system continued until September 1 when the hybrids were switched exclusively to the floating ration.

Daily minimum and maximum water temperatures were recorded from a Heath Kit Digital Weather Computer. Dissolved oxygen levels were measured with a Yellow Springs Instrument Dissolved Oxygen Meter and recorded daily.

Total alkalinity, carbon dioxide, pH, total and un-ionized ammonia and nitrite levels were taken and recorded weekly in raceways using a Hach Portable Water Quality Kit.

Disease identification was performed on-site with the exception of bacterial diseases. Suspected cases were confirmed by the Fish Farming Experimental Station, Stuttgart, Arkansas. The following methods for disease treatment were used in administering chemicals to raceways: (1) static with aeration; (2) flush; (3) oral.

Hybrids in each raceway section were sampled

periodically throughout the study. Fish were anesthetized with MS222, measured and weighed to determine growth rates and general condition of the fish.

Fish in cages and raceways were harvested manually with dip nets and a wire mesh crowder. Feeding was discontinued for two days prior to harvesting to allow accurate measurement of fish production.

Results and Discussion

Hybrids were harvested from raceways on September 16. A total of 5,773 fish weighing 658 lb were removed from the four raceways.

Average length and weight, food conversion and survival are presented in Table 2. Treatment 2 (high stocking density) contained 321 lb at harvest compared to Treatment 1 (low stocking density) which averaged 112 lb. Average length and weight were quite similar for both treatments. Treatment 2 had more efficient food conversion, 1.70, compared to 2.16 in Treatment 1. Survival was enhanced in Treatment 2, 84.6%, compared to 68.7% for Treatment 1. Data also indicate that length and weight growth rates were similar between treatments (Fig. 1).

No difference in production or water quality was apparent between upper and lower raceway sections within each treatment. Past work at Gallatin with channel catfish culture at densities up to 10 lb/ft^3 showed a decline in production toward the raceway discharge due to progressively deteriorating water quality (Collins et al. 1982a). Thus, the culture system's carrying capacity for hybrid bass is apparently much greater than the stocking densities tested in this study (maximum of 0.5 lb/ft^3). The improved food conversion and higher survival in Treatment 2 indicated that fish were more efficient at the high density. The above observations demonstrated that hybrid bass culture has the potential for much higher densities than were realized in this study.

Periodic sampling from June 22 to August 18 was conducted by a graduate student assigned to the

Table 1. Raceway stocking of white bass X striped bass--per raceway average.

Treatment	No. fish	No./ft^3	Weight (lb)	Fish/lb	lb/1,000 fish
1	1,570	2.6	5.0	314	3.2
2	3,140	5.1	8.0	393	2.5

Table 2. Total number and weight, average size, food conversion and survival of white bass X striped bass in raceways--by treatment.

Treatment	No. Harvested	Total Weight (lb)	Average Weight (lb)	Average Length (in)	Food Conversion	Percent Survival
1	1,039	112.0	0.11	6.0	2.16	68.7
2	2,655	321.0	0.12	6.2	1.70	84.6

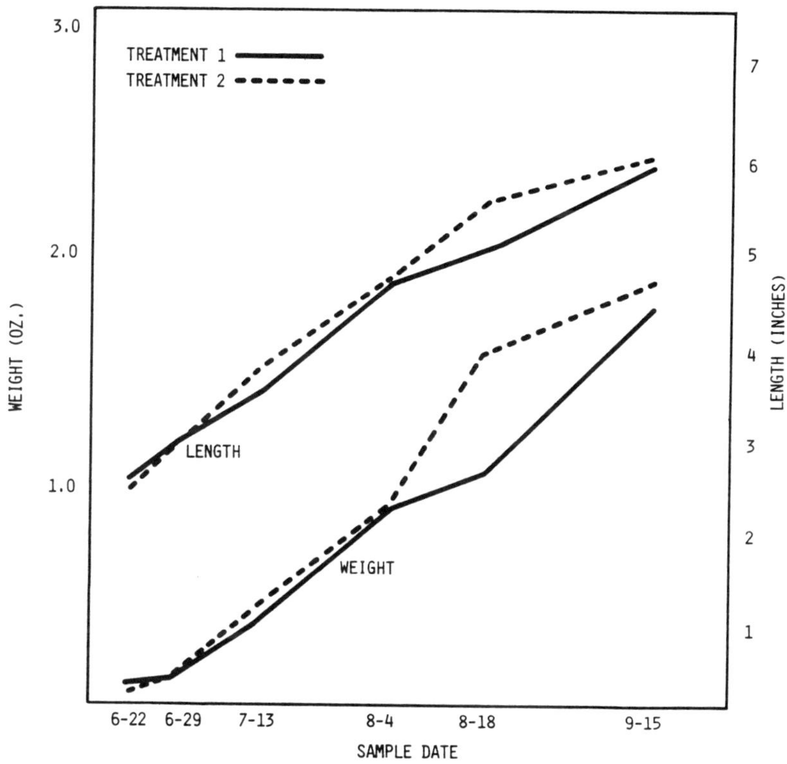

Figure 1. Length and weight of white bass X striped bass by sampling.

project to evaluate growth parameters as partial requirement for a thesis. During sampling, extreme care was exercised to keep stress at a minimum; however, more than 45 percent of the recorded mortalities in raceways occurred within three days following sampling.

Fish were fed four times daily after stocking in cages. It was apparent after several days that more frequent feedings were required due to fish feeding intensity and observed cannibalism. A marked decrease in mortality was noted after increasing the feeding rate to nine times daily and changing feed from 35 percent protein sinking trout ration to 50 percent protein trout starter ration. An automatic feeder (Falls 1980) was installed over one cage to observe fish feeding activity with a continuously available food supply. Nearly all food offered by hand and automatic feeder was consumed with negligible waste. It would probably be beneficial to the fish and culturist alike to utilize automatic feeders due to increased growth rate, greater availability of food, likely decrease in cannibalism and decreased labor costs.

Food size was increased and feeding frequency decreased as fish increased in size. When the hybrids reached four inches in total length, a floating pellet was offered periodically to evaluate acceptance. Larger fish immediately utilized the floating pellet. Size 4 sinking ration continued to be fed as a supplement, but the hybrids fed more agressively on floating feed. After September 1 fish were switched totally to a floating ration.

Hybrids fed intensely at all temperatures throughout the study. The reciprocal hybrids cultured at Gallatin in 1981 responded in a like manner (Collins, et al. 1982b). Woiwode and Adelman (1983) found that hybrid white X striped bass grew at temperatures from 51.8 to 95°F and had an optimum growth temperature of 87.8°F.

High density culture of hybrid bass provides an ideal environment for opportunistic pathogens. External parasite infestations encountered during the study period included _Trichophrya_, _Costia_, _Epistylis_,

Chilodonella and *Trichodina*. The most abundant and persistent was *Trichophrya*. All parasites were successfully controlled with one of two standard chemical treatments: (1) 85 ppm formalin flush for 60 minutes, or (2) 2.5 ppm potassium permanganate flush for 60 minutes. Prior to each treatment fish were subjected to a 2 percent salt (NaCl) static water bath for 30 minutes. This procedure was used to remove the excess mucus exposing parasites to the treatment chemical.

Two bacterial pathogens were identified: (1) columnaris and (2) *Aeromonas hydrophila*. External columnaris was a frequent but controllable problem using the aforementioned potassium permanganate treatment. Systemic infections of columnaris occurred twice in combination with *Aeromonas hydrophila*. Both were treated with the recommended level of Terramycin mixed in feed (85g active ingredient/100 lb) for 10 days with limited or no success. Furacin was tested in subsequent bioassay studies for control of columnaris and *Aeromonas hydrophila* infections using two types of treatment. Furacin was mixed with feed at a rate of 83g active ingredient/100 lb and fed for seven days. Reduction in palatability eliminated this method as an effective treatment. Furacin was also tested as a 20 ppm static water bath for 60 minutes. Neither columnaris nor *Aeromonas hydrophila* were eliminated but effective control was established with two treatments on successive days.

Two problems related to the culture environment were responsible for a greater part of the study's mortality than any of the mentioned pathogens: (1) gas bubble disease and (2) cannibalism.

Gas bubble disease (GBD) can be caused by supersaturation of atmospheric gases (mainly nitrogen) in heated effluent through changes in temperature and/or atmospheric pressure during the condenser cooling process (D'Aoust and Clark 1980). Obvious clinical signs of GBD (popeye, fin and subcutaneous bubbles, gas emboli in gills, etc.) began to appear in the second week of cage culture. Approximately 40 percent of the mortality during the cage culture phase can be attributed to GBD. Aerating incoming water (Fig. 2) successfully degassed culture water to the point that GBD was practically eliminated.

Cannibalism was evident throughout the study in daily observations and stomach analysis, but mainly during the first month of culture. As previously discussed, feeding frequency and protein level in feed were increased, which improved growth performance, satisfied appetite and reduced cannibalism. After the hybrids were stocked in raceways, observable instances of cannibalism dropped, especially in the high density groups as evidenced by an increased survival rate. Overall, from stocking to harvest, 62 percent of the total number stocked were unaccounted for. This discrepancy may be attributed to cannibalism and/or subsampling error prior to stocking. The degree of cannibalism that occurred was impossible to quantify.

Cannibalism is an inherent risk in intensive culture of carnivorous fish. However, it is apparent from the data presented that several management practices can reduce the incidence of cannibalism. Particularly important in the fry and small fingerling stages is constant availability of a high quality palatable food. Automatic feeders would be an advantage during this growth phase. Periodic grading and removal of larger individuals would preserve greater size uniformity. Comparison of survival between the two stocking densities indicated that higher densities suppressed cannibalism.

Water quality parameters shown in Table 3 were evaluated weekly, with the exception of temperature and dissolved oxygen, which were monitored daily.

Minimum and maximum water temperatures are presented in Fig. 3 showing an average Δ T of 5°F and average daily temperature of 79°F for the production period. It was noted that hybrid bass performed well at all temperatures.

Average weekly dissolved oxygen levels remained within 0.5 ppm of saturation and averaged 7.5 ppm in both treatments.

Seasonal high, low and average water quality values for alkalinity, carbon dioxide, pH, total

Table 3. Water quality - white bass X striped bass.

Range	Total Alkalinity (ppm)	Carbon Dioxide (ppm)	pH	Total Ammonia (ppm)	Un-ionized Ammonia (ppm)	Nitrite (ppm)	Temperature °F	Dissolved Oxygen (ppm)
Low	57.9	5.0	7.2	0.30	0.02	0	72.0	6.8
High	77.0	15.0	8.7	0.50	0.10	0	85.0	8.2
Average	62.9	12.9	7.8	0.40	0.06	0	79.0	7.5

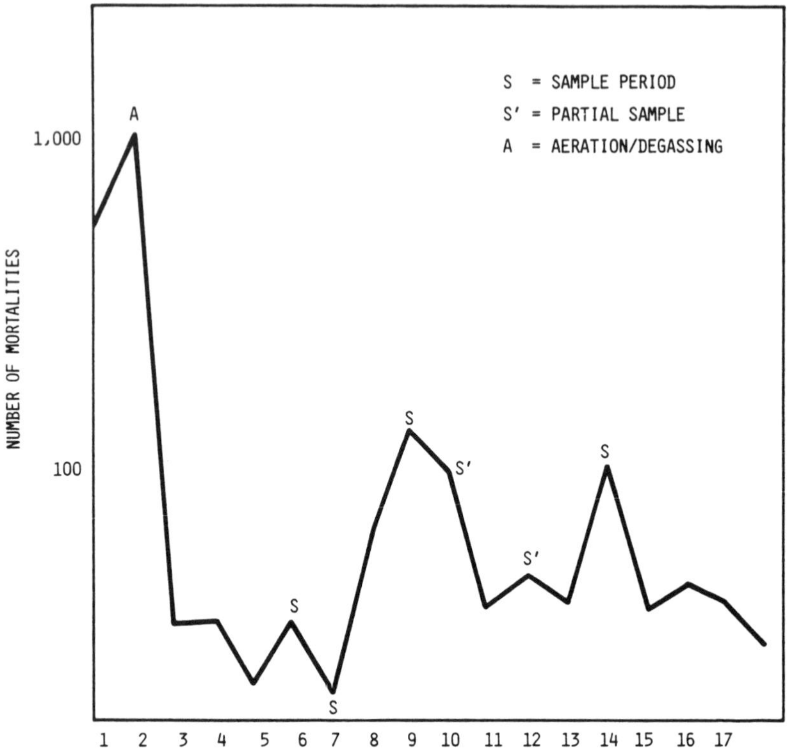

Figure 2. Weekly total mortality white bass X striped bass--May 19 to September 16.

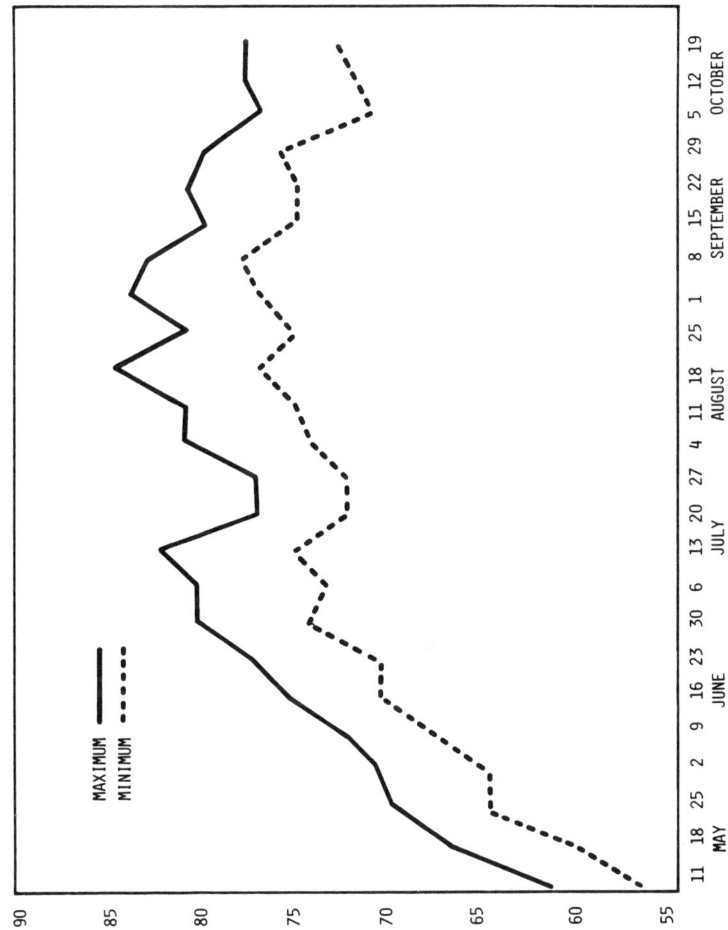

Figure 3. Weekly mean minimum and maximum water temperatures--white bass x striped bass.

ammonia, un-ionized ammonia and nitrites (Table 3) were well within limits required for hybrid bass culture. Stocking density had very little impact on water quality along the length of the raceway in either treatment.

CONCLUSIONS

1. White bass X striped bass fingerlings can be cultured at high densities (at least 0.5 lb/ft^3) without affecting performance. Production parameters were enhanced at high stocking densities.

2. Commercially available trout diets sustained acceptable growth.

3. Sampling has an adverse affect on survival and production.

4. Fish performed well at a variety of temperatures.

5. Diseases can be controlled using chemotherapeutic agents.

6. Cannibalism can be supressed with high stocking densities. Periodic grading and the use of automatic feeders should also be beneficial.

7. Water quality indicated that hybrids can be cultured at higher stocking densities than those tested.

8. Economic potential is great.

This study was made possible by the Tennessee Valley Authority (TVA) and James W. Kahrs, Osage Catfisheries, Inc.

REFERENCES

BISHOP, R. D. 1967. Evaluation of the striped bass, *Roccus saxatilis*, and white bass, *R. chrysops*, hybrids after two years. Proc. Ann. Conf. S.E. Assoc. Game and Fish Commissioners. 21:245-254.

COLLINS, C. M., G. L. BURTON and R. L. SCHWEINFORTH 1982a. Intensive culture of channel catfish in raceways utilizing power plant waste heat. Proc. Ann. CFA Research Workshop. 4:37-38.

________1982b. Raceway culture of striped bass X white bass fingerlings utilizing power plant waste heat. Proc. Ann. CFA Research Workshop. 4:39-40.

D'AOUST, B. G. and M. J. R. CLARK. 1980. Analysis of supersaturated air in natural waters and reservoirs. Trans. Amer. Fish. Soc. 109(6):708-724.

FALLS, W. W. 1980. Economical, automatic, dry-food fish feeder. Prog. Fish-Cult. 42(4):240.

LOGAN, H. J. 1967. Comparison of growth and survival rates of striped bass and striped bass X white bass hybrids under controlled environments. Proc. Ann. Conf. S.E. Assoc. Game and Fish Commissioners. 21:260-263.

STEVENS, R. E. 1964. A final report on the use of hormones to ovulate striped bass, *Roccus saxatilis* Walbaum. Proc. Ann. Conf. S.E. Assoc. Game and Fish Commissioners. 18:525-538.

________ 1983. Historical overview of striped bass culture and management. Presented at the Striped Bass and Striped Bass Hybrid Conf., Annapolis, MD, January 14, 1983. 19 pp.

WOIWODE, J. G. and I. R. ADELMAN. 1983. Growth, food conversion efficiency, and survival of the hybrid white X striped bass as a function of temperature. Presented at the Striped Bass and Striped Bass Hybrid Conf., Annapolis, MD, January 14, 1983. 10 pp.

WOODS III, L. C., J. H. KERBY and M. T. HUISH. 1983. Estuarine cage culture of hybrid striped bass. Abstracts World Mariculture Soc. Tech. Sessions, Aq. '83, Washington, D. C., January 9-13, 1983. 80.

Growth, Food Conversion Efficiency and Survival of Hybrid White Bass x Striped Bass as a Function of Temperature

John G. Woiwode and Ira R. Adelman

Natural stocks of Atlantic coast striped bass have been declining dramatically in the past 10 years, inciting considerable interest in the the hybrid white x striped bass (*Morone chrysops* x *Morone saxatilis*) as a fish to supplement the declining populations for the sport fishery and food fish industries. There is, however, little information on the temperature requirements of the hybrid for aquacultural applications.

Cox and Coutant (1981) reported that maximum growth of juvenile striped bass fed a maximum ration occurred near 24°C. Lewis and Heidinger (1981) selected 25°C for rearing striped bass fingerlings to marketable size. Coutant (1977) reported a preferred temperature for 3 year old striped bass at 22°C, while Barans and Tubb (1973) reported a preferred temperature for young of the year white bass at 31°C during summer. However, we could find no information on temperature effects on hybrid white x striped bass growth or preferred temperature.

In an effort to assess the temperature requirements of the hybrid bass for intensive thermal aquacultural applications, experiments were conducted to: 1) define the optimum temperature for growth of the hybrid bass fed at satiation levels, 2) determine the optimum temperature for conversion efficiency at satiation rations, and 3) define the approximate upper lethal temperature of this fish in terms of a critical thermal maximum.

MATERIALS AND METHODS

Growth

One-hundred hybrid bass fingerlings, with a mean

weight of approximately 8 g were shipped to the University of Minnesota fisheries laboratory from Osage Catfisheries, Osage Beach, Mo., on March 16, 1982. These fish were given a complete health examination and found to be suffering from an infestation of the ciliate protozoan Chilodonella sp. A 12 minute dip in a 1% NaCl solution, followed by a holding period in a solution of 0.5% NaCl plus 30 mg/L oxytetracycline, diluted to zero concentration after 30 minutes, cleansed the fish of the parasite.

The fish, intially held at 17°C, were randomized into seven groups of 11 fish each. From March 16 through April 12, these groups were gradually acclimated to final culture temperatures of 11, 15, 19, 23, 27, 31, and 35°C. At the end of the acclimation period, the fish were individually cold branded (liquid nitrogen), dorsal to the lateral line, and initial lengths (nearest 1 mm) and weights (nearest 0.1 g) were determined.

Well water (hardness of approximately 250 mg/L $CaCO_3$) was maintained at the desired temperature by thermostatically controlled heat exchange systems in a constant head water supply tank for each temperature. The fish were cultured in 130 L nalgene tanks measuring 60x45x45 cm. Water inflow of 4 L/min, provided a volumetric changeover rate of 2.2 per hour. The low velocity, high turnover design seemed suited to the bass' environmental requirements, although no experiments were performed to optimize the water-container interaction.

Although the hybrid bass accepted hard pellets, they would often roll them around in their mouths and ultimately spit them back out with many pellets going to waste. This problem of acceptance seemed to be due to food texture. To assure a satiation level of feeding with no waste, the food had to be presented in a form that resulted in 100% consumption. A diet of 42% Silver Cup Salmon Starter Mash, 7.5% Knox's unflavored gelatin, and 50.5% water, pelletized into a long, soft pellet proved satisfactory.

The groups of fish at each temperature were cultured for 30 days, with one growth assessment after

15 days. For each group of fish daily weight gain was estimated on a logarithomic growth curve. Food was fed to achieve the projected weight gain at an estimated conversion efficiency. The estimates were assessed daily on the basis of whether or not fish actually consumed the entire day's ration. If the estimate for a given day was low, additional food was provided. Food was pre-weighed and hand-fed six times a day. Uneaten food was negligible at 11°C and non-existent at other temperatures. No attempt was made to recover this food at 11°C.

Specific growth rate of each fish was calculated as:

$$G = \frac{\log_n W_f - \log_n W_o}{\text{days}}$$

where W_f is the final weight and W_o the initial weight. Conversion efficiency (%) for each tank of fish was calculated as:

$$CE = \frac{W}{C} \times 100$$

where W is the total wet weight gained by the fish and C is the total wet weight of food offered.

Critical Thermal Maximum

The upper lethal temperature of the hybrid bass was determined as a critical thermal maximum (CTM) (Becker and Genoway 1979). Ten fish were acclimated to 34.5°C for a period of 30 days prior to the experiment. Weight of fish at the time of testing ranged from 23 to 41 g. The fish were individually placed in a 4000 ml beaker 3/4 filled with water at 34.5°C. The water was then immediately heated at the rate of 0.3° C min^{-1}. The temperature was noted when fish lost equilibrium after which they were immediately returned to water at their acclimation temperature.

RESULTS AND DISCUSSION

Growth

Growth occurred at all test temperatures (Fig. 1). Although the fish were cultured over a wide range of temperatures (11-35°C), zero growth rate did not occur. The eurythermic response of the hybrid is especially desirable for aquaculture applications where feeding and growth will occur over a wide range of temperatures. The optimum temperature for growth was 31°C although the specific growth rate at 27°C was almost as great (Fig. 1, Table 1). Total food consumed corresponded directly to growth, peaking at 31°C and declining sharply at 35°C (Table 1). In spite of the reduction in growth and food consumption at 35°C, the fish still grew well at this high temperature with a mean specific growth rate of 2.6% per day. Cox and Coutant (1981) noted zero growth in juvenile striped bass at 33.5°C.

Peak conversion efficiency occurred between 19 and 23°C, and slowly declined as temperature increased, with a sharp decrease in efficiency at 35°C (Fig. 1). The 12°C difference between the optimum temperature for growth and peak temperature for conversion efficiency appears somewhat unusual. Coutant and Cox (1981) found peak conversion efficiency and growth rate for striped bass to be at the same temperature. The satiation ration, adjusted daily in the present study, was probably a relatively higher level than the 100% ration in the striped bass study. The decline in conversion efficiency at temperatures above 19°C probably reflects a large metabolic energy demand at the higher temperature in combination with an apparently higher specific dynamic action (SDA) associated with the greatly increased food consumption. Temperatures of 27 and 31°C resulted in dramatically increased food consumption and significantly greater growth, but at a higher bioenergetic cost than at 19 and 23°C. Determination of the maintenance ration at the different temperatures, as well as the effects of different ration sizes on growth and conversion efficiency

Table 1. Mean initial and final length and weight, specific growth rate, food consumption, and food conversion efficiency of white x striped bass hybrids cultured at various temperatures. (S.D. in parenthesis, N=11.)

Temperature (°C)	Initial length (mm)	Final length (mm)	Initial weight (g)	Final weight (g)	Specific growth rate (% day^{-1})	Food consumed (g)	Conversion efficiency (%)
11	92 (10)	97 (12)	7.6 (0.8)	9.3 (1.0)	0.60 (0.11)	59	27
15	96 (12)	109 (12)	9.3 (1.2)	14.1 (1.6)	1.46 (0.28)	144	36
19	100 (14)	121 (14)	10.1 (1.3)	20.4 (2.3)	2.41 (0.25)	220	51
23	98 (8)	132 (8)	9.3 (0.7)	28.2 (1.7)	3.75 (0.33)	413	50
27	95 (12)	140 (12)	8.7 (1.0)	36.0 (3.2)	4.81 (0.32)	623	48
31	94 (8)	138 (10)	8.3 (0.7)	38.6 (2.8)	5.13 (0.27)	763	44
35	93 (14)	107 (13)	8.7 (1.3)	18.4 (2.1)	2.60 (0.37)	362	29

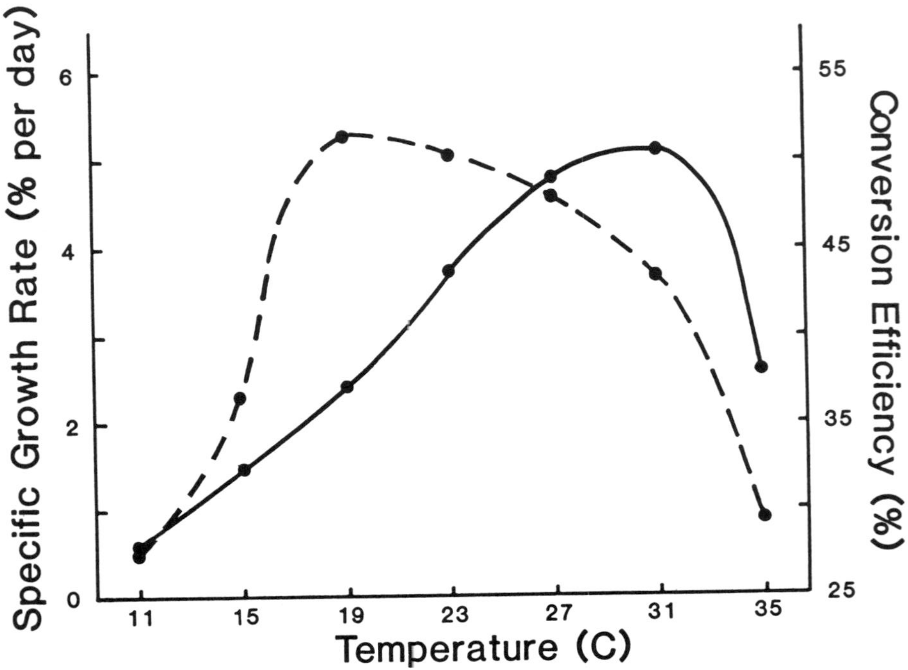

Figure 1. Specific growth rate (solid line) and food conversion efficiency (dashed line) at various temperatures for hybrid white x striped bass fed a satiation ration for 30 days.

at all temperatures will provide a more complete understanding of the bioenergetics of the hybrid.

Peak conversion efficiencies of hybrid bass in this study were in the range of 50% (Table 1), based on the wet weight of the food containing approximately 50% water. If conversion efficiencies were based on dry weight of food and wet weight of fish, peak efficiencies would appear to be close to 100%.

Critical Thermal Maximum

The mean CTM for the 10 hybrid bass acclimated to 34.5°C was 39.04°C (± 0.53 S.D.). All but one fish ultimately died following the CTM test, even though the fish were immediately returned to water at their acclimation temperature upon loss of equilibrium. Loss of equilibrium, rather than death, was used as an endpoint because it could be observed more precisely and because it is as meaningful an ecological endpoint as death.

This CTM value is probably slightly lower than might have been attained if the hybrid bass were acclimated to a higher temperature. The latter certainly seems possible since the bass in the growth study acclimated to 35°C still had a specific growth rate of 2.6% per day. It seems that acclimation to a higher temperature where growth rate would be zero is possible. A higher CTM value for fish at that acclimation temperature is likely.

Although these experiments are preliminary, they provide some valuable insights into the intensive culture of this fish. Optimum temperature for growth of the juvenile hybrid bass on a satiation diet is considerably higher than that of the striped bass juveniles reported by Cox and Coutant (1981), yet the hybrid still grows rapidly over a wide temperature range.

Funds for this study were provided in part from Northern States Power Company Aquaculture Project and University of Minnesota Agricultural Experiment Station Project 75. We thank Osage Catfisheries for supplying fish and David Mullenbach and James I. Stewart for assistance in conducting experiments.

REFERENCES

Barans, C. A. and R. A. Tubb. 1973. Temperatures selected seasonally by four fishes from Western Lake Erie. J. Fish. Res. Board Can. 30: 1697-1703.

Becker, C. D. and R. G. Genoway. 1979. Evaluation of the critical thermal maximum for determining thermal tolerance of freshwater fish. Env. Biol. Fish. 4: 245-256.

Coutant, C. C. 1977. Compilation of temperature preference data. J. Fish. Res. Board Can. 34: 739-745.

Cox, D. K. and C. C. Coutant. 1981. Growth dynamics of juvenile striped bass as functions of temperature and ration. Trans. Amer. Fish. Soc. 110: 226-238.

Lewis, W. M. and R. C. Heidinger. 1981. Tank culture of striped bass production manual. Fisheries Research Laboratory, Southern Illinois University, Carbondale, IL. 115 pp.

Striped Bass Feeds and Feeding

T.R. Zeigler, L.C. Woods, III,
and J. Gabaudan

In order to more fully appreciate the concept of feeds and feeding for striped bass a brief introduction concerning the scope and impact of this subject is presented.

First, the objective should be clearly identified. What is the purpose of the culture operation and why are the fish being fed? Brood fish culture with subsequent egg production may be desired. Larval, fry or fingerlings are desired in large numbers. Adults may be grown for food, sport fishing or other recreational purposes. Commercial operations would differ from research. The objective may be targeted towards large numbers or maximum pounds. The rearing system may be intensive or extensive. Clearly, not one feed or one feeding program will likely meet the needs of all of the above objectives. The objective requires careful identification and then must be matched with the best available or customer designed feed and feeding program.

Second, feed is a primary cause of the decline in water quality. Oxygen is used up as the consumed food is metabolized. Various end products are produced. Fines or uneaten feed negatively impact water quality. Management of the feed or feed program can significantly alter water quality.

Third, feed can also be used to supply non nutrient factors. It can serve as a vehicle for medication, vaccines or perhaps other additives such as pigmenting agents or compounds used to aid in influencing environmental factors such as water quality and environmental organisms.

Fourth, scientific facts from research concerning the subject of feeds and feeding and the nutrition of striped bass are sparse. Most meaningful information is to be gained from the analysis and interpretation of large scale operations which have a history of successful results.

Fifth, there are many significant non feed variables inherent to nearly all aquaculture systems that can significantly affect the results of any feed or feed program. With more complete knowledge or control of these variables, it becomes easier to formulate diets for optimum efficiency, i.e., diets that are best targeted to the specific production system being used.

Therefore, for the purpose of this presentation, we will emphasize certain basic principles in aquaculture nutrition. The physiology and biology of the striped bass will be examined and compared to other physiologically similar species which have been studied more extensively. From these comparisons, correlations can be cautiously drawn which suggest opportunities for more effective feeds and feeding programs for the striped bass. We will also discuss commercially available products and conclude with a general discussion of suggested feeding procedures and guidelines.

INTERPRETING PHYSIOLOGY AND BIOLOGY

In those situations where the nutritionist is called upon to make nutritional and feeding recommendations for species for which little scientific evidence has accumulated one is able to make intelligent estimates by examining the normal physiology and biology of the animal and extrapolating this information to some species where more complete nutritional information is available.

The striped bass is an adaptable species of anadromous fish found along most of the sea coast of the United States. Striped bass are carnivorous which is reflected by the structure and function of the alimentary canal. It consists of a large mouth with teeth, short esophagus, large stomach and a relatively short intestinal section. The rate of food passage has been determined to be from 1 1/2 to 5 hours depending on the age of the fish. Extensive studies reveal that natural food consists of a wide variety of zooplankton, insects, crustaceans and other fish. The particular type of food consumed at any one time depends largely upon the type of food available, the size of the fish and the size of food. Vegetative material is not found in the intestinal tracts of wild fish. This digestive physiology and these food habits suggest that striped bass would be good utilizers of dietary fat and protein and poor utilizers of raw carbohydrates and fiber. Activity and growth rate are temperature dependent, which is typical of cold blooded species. A study of the growth dynamics of juvenile striped bass by Cox and Coulant (1981) determined that maximum growth occurred near 24°C with zero growth occurring at the extremes of approximately 10°C and 35°C. The young hatch from eggs as larvae and develop into fingerlings at about 30-50 days of age. The growth of the fish continues at a relatively constant rate to four years of age and then continues at a slower rate for up to 14 years (Lewis and Heidinger, 1981). The rate of growth is dependent upon environmental temperatures and food supply. During larval and early stages the rate of growth is very rapid, doubling body weight every 5-10 days. This suggests frequent feeding of high energy and highly digestible feeds will be essential for effective culture at this stage.

Based on this information, it is suggested that the nutritional requirements of striped bass will be most similar to the nutritional requirements of trout and salmon adjusted for the difference between optimum temperatures, 20°C for striped bass and 14°C for rainbow trout. The second significant difference is the larval stage for the striped bass which does not occur in the trout or salmon. It would be extremely interesting to compare the growth rates of trout and striped bass after the fingerling stage to determine if the differences in growth rates could be accounted for by the 10°C difference in optimum temperature.

Growth curves, weight plotted against time, represent the response to environment. They are useful in predicting. Energy and nutrient requirements are a function of weight. If the weight is known the feed requirements can be predicted. Rate of gain can be used as a standard to compare fish performance from year to year or among culture systems.

ENERGY AND ENERGY METABOLISM

In order to precisely formulate feeds, the energy requirements of the fish and the energy contribution of the ingredients used in the feed must be known. Animals, including fish, normally adjust their feed intake to meet their energy requirements. Therefore, a basic understanding of energy metabolism and energy measurement is essential for even the most basic understanding of feeds and feeding programs.

Fish require energy for growth, activity and reproduction. This energy is derived from the oxidation of the food. The biological process of utilizing energy is defined as metabolism and the rate at which energy utilization occurs is called metabolic rate. The metabolic rate or energy requirements are influenced by many factors including temperature, body size, activity and water chemistry.

Fish utilize the dietary components of fat, protein and carbohydrates differently from other animals. The utilization of the components especially carbohydrate also varies among species of fish.

Energy values are measured and expressed in terms of calories, whether it be the energy requirement of the animal or the energy contribution of the feed or feed ingredient. Potential confusion concerning the understanding of energy metabolism begins at this point. Several different measures of food energy are used to define the energy requirements (Table 1). Some measures are direct and very scientific, others are indirect, and some are basically calculated estimates based on energy measurements of a variety of species. Therefore, energy values may be measured and expressed in terms of gross energy, digestible energy, metabolizable energy, net energy or physiological fuel values. Frequently in the recording of these values, the identification or definition of the values is not given. This situation severely clouds interpretation of energy data in scientific literature. Easy to understand discussions concern-

Table 1. Energy values used to calculate the energy content of fish feeds

EXAMPLE FISH FEED		KCAL/GRAM GROSS 1/	P.F.V. 2/	CATFISH 3/	TROUT 4/
PROTEIN	40%	5.65	4.0	3.5	3.9
FAT	10%	9.45	9.0	8.1	8.0
NFE	27%	4.10	4.0	2.5	1.6
FIBER	3%	--	--		
ASH	10%	--	--		
MOISTURE	10%	--	--		
	100%				
		4.31	3.58	2.89	2.79
		100%	83%	67%	65%

1/ GROSS ENERGY VALUES BY BOMB CALORIMETER

2/ PHYSIOLOGICAL FUEL VALUES, MAYNARD & LOOSLI 1969

3/ NRC WARMWATER FISHES 1977

4/ NRC COLDWATER FISHES 1981

ing the energy concept in fishes are given in the NRC publications on the nutritional requirements of cold water and warm water fishes.

In fish the energy requirement or metabolic rate is dependent on water temperature. The law of physical chemistry which states that the rate of chemical reaction doubles or halves for every 10°C change in temperature is closely approximated in the data for fishes. Therefore as the temperature changes, the feeding rate or energy concentration of the feed must be changed if the energy requirements of the fish are to be met. Lack of attention to this very important principle can easily result in under feeding or over feeding.

The energy required for activity constitutes a significant portion of the energy required for the basic metabolism of the fish. Activity increases with rising temperature. This activity is due to increased feeding activity and activities related to social structure. Management systems controlling activity will enjoy a greater percentage of food energy channeled towards weight gains as opposed to body maintenance. Historically, feed energy has been calculated using various energy values for the feed components of protein, fat and carbohydrates. Feed energy values calculated in this way may vary considerably.

A further problem in measuring and managing energy requirements of fish through fish feed is that energy values of feed ingredients have not been determined to a significant degree of completeness. Digestible energy and metabolizable energy values have recently been reported for approximately 40 ingredients for rainbow trout; but whether these values can be used in the formulation of feeds for striped bass remains unclear. Although these values may not be precise for striped bass, they may be accurate on a relative basis. Energy values for catfish and other warm water species are quite limited and the best available information concerning energy utilization is published in the form of digestion coefficients.

Good information concerning the energy concentration of ingredients is absolutely basic to the scientific formulation of feeds. The underlying principle is to first meet the energy requirements of the animal through that quantity of feed which will be consumed and then balance all of the other nutrients, protein, amino acids, vitamins and minerals in the diet so that their relative proportions meet the requirements of the animal. Nutritional requirements are most precisely expressed in terms of milligrams of nutrient per unit of body weight per day under defined environmental conditions. These nutrients must be formulated into the diet at those proportions which will supply the daily requirements on a continuous ongoing basis.

PROTEIN AND AMINO ACIDS

Little has been published concerning the protein and amino acid requirements of striped bass. Millikin (1982) fed graded levels 34%, 44% or 55% crude protein to 2 1/2 gram fish for a 6 week study. He concluded that the highest protein level produced the best weight gain and feed efficiency but that protein utilization was highest for the 34% protein diet. Millikin's diets were made isocaloric based on the physiological fuel values for energy of Maynard and Loosli (1969).

Since little else is known from scientific literature, it is not possible to present scientific evidence concerning protein requirements. However, it is possible to examine the concept of protein requirements; and if certain facts are known or can be predicted, the protein requirement can be estimated by calculation.

Dietary protein is primarily used for the synthesis of tissue protein. Targeting on tissue protein, we can then conclude that an increase can only result from biologically available food protein. The one must equal the other. Increased tissue protein is a function of weight gain calculated using the percent protein content of the whole carcass. The food protein available for protein synthesis is equal to the food consumed times the percent protein in food times the biological value of the food protein.

Food consumed is estimated by multiplying the weight gain times the food conversion expressed as a ratio of food consumed over weight gained. Using these factors, an equation is produced predicting a protein requirement expressed as percent of protein in the feed (Table 2).

Cowey and Sargent (1972) in their review of fish nutrition report food protein biological values ranging from 20% to 45% for trout. Millikin reported values for striped bass ranging from 23% to 34%. It is apparent that considerable variability exists with this parameter.

If we assume the protein content of striped bass as 20% and a feed conversion of 1.7 for adult fish and a biological value of .3, the protein requirement is then calculated to be 39.2%.

It should be emphasized that there is considerable confusion in the reported literature regarding interpretation of growth and diets fed. Unless carefully formulated, increasing the protein content of the diet may also be increasing the energy content. It is well documented in many animals that fat spares protein. The lack of accurate energy data for feed ingredients for various species of fish further complicates the problem. Nutritional studies reporting protein or energy effects need to be carefully reevaluated utilizing the most recent scientific facts concerning protein utilization and the most precise energy values.

A review of NRC publications reveals that most aquatic species for which good amino acid data have been reported have amino acid requirements which are quite similar. The amino acid requirements of trout have been reasonably well determined and since the physiology of the two species are similar we would suggest that the amino acid requirements for salmonids should closely approximate the amino acid requirements of striped bass and could be used until further evidence is reported.

FEEDS

The type and formula of feeds required in the rearing of striped bass will depend on the culture system

Table 2. Estimating required level of dietary protein

		BASIC EQUATION
INCREASE IN TISSUE PROTEIN	=	BIOLOGICALLY AVAILABLE PROTEIN FROM FEED
WT GAIN X % TISSUE PROTEIN	=	FOOD CONSUMED X % FEED PROTEIN X BIOLOGICAL VALUE
	=	WT GAIN X FOOD CONVERSION X % FEED PROTEIN X BIOLOGICAL VALUE
WG X % TP	=	WG X FC X % FP X BV

$$\frac{\text{WG X \% TP}}{\text{WG X FC X BV}} = \text{\% FEED PROTEIN}$$

$$\frac{\text{\% TP}}{\text{FC X BV}} = \text{PROTEIN REQUIREMENT AS \% PROTEIN IN FEED}$$

$$\frac{20}{1.7 \text{ X } .3} = 39.2\text{\% PROTEIN}$$

being used. Where extensive culture is practiced, the feed normally contributes supplemental energy, minerals and some protein to the natural food supply. In ponds it may serve as a substitute for the production of natural pond food. For intensive culture conditions the feed of course must be complete and balanced and contribute all the essential nutrients for the fish at the designated stage in the life cycle.

Since the aquaculture industry is quite small compared to other animal production industries, there are relatively few companies manufacturing fish feeds. Those making any significant amount of fish foods number probably less than 15; and one could count on one hand the number of companies manufacturing a more complete line of aquaculture diets which have available the technical and physical resources to meet the needs of a young industry.

Production of aquaculture feeds requires special milling equipment and manufacturing techniques. Unlike other animal foods such factors as palatability, texture, water stability and high fat levels are important in aquaculture diets. The result is that the fish culturist does not have readily available feeds that are ideally suited and many times has to pretty much take what is available if the fish are to be fed. This is especially true when raising fish that have not been extensively commercialized, unlike trout and catfish culture.

Fish feeds are unique in several other ways. They normally contain higher levels of fihsmeal and fish oils, which are subject to more rapid oxidation unless properly protected by suitable anti-oxidants. Vitamin C is required by fish and not by most other animals. Vitamin C is very unstable. In the manufacturing operation, 15% to 50% of the added Vitamin C may be destroyed due to the manufacturing process. Destruction during storage may amount to 15% to 30% per month. Although some attempts have been made to stabilize Vitamin C, we cannot yet conclude that feeds can be fortified with a stabilized Vitamin C that has been proven biologically available to all aquatic species.

A brief discussion of the more commonly available

types of feed which could be used in striped bass culture are shown in Table 3 and discussed below.

Brine shrimp. Brine shrimp nauplii are routinely fed to striped bass larvae as the first feed fed after hatching. In recent years significant advancements have been made in the quality of brine shrimp. Brine shrimp composition is influenced by the environment in which they are raised and will tend to concentrate toxic contaminats if present. Recent research has indicated that the essential fatty acid profile also varies in brine shrimp from different locations. There is also variation in the size of the hatched nauplii depending on the size of the egg.

Single cell proteins. A wide variety of single cell bacteria and yeast have been produced which have been used as ingredients in animal feeds. Perhaps the most common of these are dried brewers yeast and dried bakers yeast. Using tools such as genetic engineering, it is possible that in the future strains of yeast could be developed that would function as complete diets for larval species.

Microencapsulation. This new technology also offers promise in manufacturing larval diets. The objective would be to prepare a media that was nutritionally complete, microencapsulate the media with a water resistant coating that contains attractants. These microcapsules could be preduced in a variety of sizes.

Flaked or larval diet. These diets represent the most successful prepared food which is being fed within several days of hatching as a supplement to brine shrimp. They are produced by blending various high quality ingredients with various vitamins, minerals and special supplements and then homogenizing and mixing with water to form a puree and drying them on a steam heated drum dryer. The resulting product is sheets of dried diet approximately 75 microns thick. The product is then ground and sieved to produce fine granules of a variety of sizes.

Table 3. Examples of commercially available feeds

	PROTEIN %	FAT %
BRINE SHRIMP, DRY WT.	60-65	6-7
SINGLE CELL PROTEIN (SCP)	40-65	4-8
LARVAL DIETS	40-55	6-12
SALMON STARTER	48-55	13-17
NO. 1 CRUMBLES	48-55	13-17
NO. 2 CRUMBLES	48-55	13-17
NO. 3 CRUMBLES	43-50	12-16
NO. 4 CRUMBLES	43-50	12-16
SALMON GROWER		
3/32" PELLETS	40-48	10-15
1/8" PELLETS	40-48	10-15
5/32" PELLETS	40-48	10-15
3/16" PELLETS	40-48	10-15
TROUT STARTER	40-45	8-12
TROUT GROWER	34-40	6-13
CATFISH GROWER	25-35	5-8
POND FEEDS	20-36	4-6

FEED COMPOSITION AND SIZES VARY AMONG MANUFACTURERS

Salmon diets. Salmon diets are the most highly concentrated diets commercially available. They contain high levels of protein and fat along with high levels of complex vitamin mixes and selected minerals. They normally contain high levels of fish protein and fish oils.

Trout diets. Trout feeds are formulated very similarly to salmon feeds but frequently do not contain as high a nutrient density. Because trout are not picky eaters, nutritionists will change formulations to meet specific nutritional standards based on price and availability of ingredients.

Catfish feeds. The nutritional requirements for catfish are not as high as those for trout and salmon and accordingly the diets normally contain lower levels of protein, fat and vitamins. Since most catfish are pond raised and capable of getting some natural food from the pond, many catfish feeds serve as supplements as opposed to being a complete ration. Catfish feeds also contain higher levels of raw carbohydrates.

Pond feeds. Pond feeds typically contain the lowest nutrient levels of commercially available aquaculture diets. They may contain high levels of fiber which serves as an organic substrate for pond organisms. Having low levels of protein and normally lower levels of vitamins, they serve as a supplement feed to the pond fish who balance their ration from other sources of food.

Meal. Those diets which are in meal form are a dry blend of ingredients and additives with or without grinding. Water soluble components are rapidly lost in the water environment upon feeding.

Pellets and crumbles. Most fish feeds are pelleted. This process involves dry blending the ingredients, pre-conditioning with steam whereby the moisture is increased by 3% and the temperature is increased to approximately 70-80°C after which the hot moist feed is forced through holes in a die, cut, cooled and sacked. Crumbles are manufactured by crushing the pellets and sieving to certain standard sized particles.

Extruded feeds. Extruded feeds are often called floating feed as this process produces nuggets that are catacombed with air spaces and consequently float when placed in water. These feeds are manufactured by dry blending the ingredients, adding 10% to 15% water, pressure cooking at temperatures up to 125-155°C and pressure extruding them through orifices. The expansion occurs when the hot moist food under pressure is exposed to atmospheric conditions. The nuggets are dried and sacked.

Feed sizes. As many as 15 different sizes of feed may be required to adequately grow fish from larva or fry to brood stock. Selection of the proper size feed to be fed at the proper time is an important factor in culture management. Feed sizes may vary among manufacturers. In order to obtain the best information, the technical information concerning specific products being considered should be thoroughly reviewed and discussed with the manufacturer (Table 4).

Feeds formulated according to confidential specifications are known as closed formula feeds. The US Fish and Wildlife Service purchases feed according to fixed detailed specifications. These feeds are known as open formula feeds. There are benefits to both program and the culturist should evaluate each as it relates to the proposed objectives (Tables 5, 6 & 7).

Since vitamin destruction, especially Vitamin C, occurs rapidly in fish feeds, it is important that fresh feeds are purchased and inventories rotated in order to avoid feeding marginal nutrition due to age of feed. It is further recommended that all feed be dated with an easily identifiable date. For specific shelf life guidelines manufacturers recommendations should be followed. The shelf life of feeds depends on the environmental storage conditions.

Table 4. Sizes of commercially available fish food pellets and crumbles

	U S MESH		MM	
CRUMBLES OR GRANULES[1/]	THROUGH	OVER	THROUGH	OVER
STARTER	30	40	.595	.420
NO. 1	20	30	.841	.595
NO. 2	16	20	1.19	.841
NO. 3	10	16	2.00	1.19
NO. 4	6	10	3.36	2.00
PELLETS (DIAMETER)				
3/64 "			1.2	(NOT STANDARD)
1/16 (4/64)"			1.6	(NOT STANDARD)
3/32 (6/64)"			2.4	
1/8 (8/64)"			3.2	
5/32 (10/64)"			4.0	
3/16 (12/64)"			4.8	
1/4 (16/64)"			6.4	
3/8 (24/64)"			9.5	

1/ SIZES FOR CRUMBLES ARE ACCORDING TO U.S. FISH AND WILDLIFE SPECIFICATIONS. ACTUAL SIZES MAY VARY ACCORDING TO COMPANY.

Table 5. Open formula feed specification (USFW)

SECTION 1
FORMULATION SPECIFICATIONS FOR STARTER DIET, SD9-30
(Starter, No. 1, and No. 2 granules)

Item	%
1. Fish food shall be composed of the following items. The final product shall carry the following guaranteed analysis: Crude protein, not less than 50% Fish meal protein, not less than 33% Crude fat, not less than 17% Moisture, not more than 10.0% at sack-off	
2. Fish meal: Stabilized, maximum moisture 10%, maximum salt 5%, stored at manufacturer's no longer than 6 months as indicated by the Bill of Lading. Pepsin digestibility not less than 92.5%. Different meals may not be combined for use in the feed.	Not less than 50
3. Wheat feed flour: Minimum protein 14%, maximum fiber 1.5%	10.3[1]
4. Soy flour: Defatted, minimum protein 48.5%, maximum fat 1% (flour must be adequately toasted with a protein dispersibility index of less than or equal to 20).	15
5. Dried blood flour or ring dried blood meal: minimum protein 80%.	10
6. Trace mineral premix No. 2 (see Section 5 of specifications).	1#/ton
7. Vitamin premix No. 30 (see Sections 4 & 6 of specifications).	8#/ton
8. Choline chloride, 50%	4.5#/ton
9. Ascorbic acid	1.5#/ton
10. Fish oil: stabilized with 0.04% BHA-BHT (1:1) or 0.01% ethoxyquin, less than 3% free fatty acids and must meet standards for peroxide value and TBA value established by the Diet Testing Development Center.	12[1]
11. Lignin sulphonate pellet binder (e.g. Ameribond, Orzan, or equivalent).	2

[1] Fish meal may be increased depending upon protein content but must provide not less than 33% fish protein. Quantity of added oil may be adjusted so that the finished feed shall contain not less than 17% crude fat. Wheat feed flour is to be adjusted to compensate for the above variations. Not less than 6% of the total fat shall be sprayed on the granules as a top dressing, the rest to be included in the feed mix.

Table 6. Open formula feed specification vitamin premix (USFW)

SECTION 4
SPECIFICATION FOR VITAMIN PREMIX NO. 30

Vitamin	Guaranteed potency per pound of premix (grams unless otherwise listed)
D calcium pantothenate	12.0
Pyridoxine (pyridoxine HCl)	3.5
Riboflavin	6.0
Niacinamide	25.0
Folic acid	1.0
Thiamine (thiamine mononitrate)	4.0
Biotin	40.0 mg
Vitamin B_{12}	2.5 mg
Menadione sodium bisulfite complex	1.25
Vitamin E (d or dl alpha tocopherol acetate)	40,000 i.u.
Vitamin D_3, stabilized	50,000 i.u.
Vitamin A (vitamin A palmitate or acetate), stabilized	750,000 USP

Choline chloride, ascorbic acid, and the vitamin premix No. 30 are to be stored separately and never mixed one with another before being added to the feed mixture.

The certified vitamin premix is to be supplied by a recognized manufacturer and must show the date of preparation. The vitamin premix to be used is not to be held in storage longer than 4 months after date of preparation.

The vitamin premix is to be made with a wheat or soybean by-product base. Rice hulls or oat feed are not acceptable.

Table 7. Open formula feed specification mineral premix (USFW)

SECTION 5

SPECIFICATION FOR TRACE MINERAL PREMIX NO.1

Mineral	Guaranteed Analysis of Element (g/lb. mineral mix)
Zinc sulfate ($ZnSO_4$ - 84g/lb of mineral mix)	34
Manganous sulfate ($MnSO_4$ - 25g/lb of mineral mix)	9.1
Cupric sulfate ($CuSO_4$ - 1.75g/lb of mineral mix)	0.7
Potassium iodate (KIO_3 - 0.38g/lb of mineral mix)	0.23

An inert carrier can be used to make up the mixture to the pound.

The mineral mixture is to be added at 1.0 pound per ton SD9 feed and 2.0 pounds per ton for GR3, GR4 and GR5 feeds.

FEEDING THE FISH

There is an old adage which states, "The Eyes of the Master Fatten the Calf." This saying applies, of course, in all animal operations but it is especially true in fish culture, where our knowledge of feeds and feeding is better described as art than sophisticated science.

The underlying principle that must be constantly addressed is that through the feed and feeding technique we must meet the physiological requirements of the fish directed toward the objective for which they are raised. Unquestionably, the fish themselves are the best indicators as to whether this need is being met and it is up to us to be smart enough to understand what the fish can tell us. This requires a good eye, some good records and some scientific common sense.

There are many non nutritionally related factors that can influence the outcome of a culture system (Table 8). Many problems or conditions that appear to be nutritionally related may not be factors of the feed formulation but are related to some management technique. Some of the most important factors include--temperature of the water, water flow, water chemistry, size of the fish, age of the fish, conditions (stress) of the fish, disease, feed particle size, time of day fed, frequency of feeding, distribution of food over the surface area and rate at which the fish are being fed. Aquaculture is unique in that fish culturists have to understand the relationship of these important variables to the success of the culture operation.

Because the actual feeding of the fish contributes significantly to the management of production variables there are new and renewed interests in better feeding techniques. Although not a new concept, feed delivery systems incorporating the demand feeder concept are showing exceptional promise. The primary benefits of demand feeders include increased growth rates from 15% to 30% and an increase in feed conversions of up to 15%. There is less feeding activity and water quality is improved. Fish health

Table 8. Considerations for effective feeding

- TEMPERATURE
- GROWTH RATE
- ACTIVITY
- SIZE
- FEED SIZE
- DISTRIBUTION
- FREQUENCY
- RATE OF DELIVERY
- STOCKING DENSITY
- WATER QUALITY
- DISEASE
- LIGHT
- WEATHER CONDITIONS
- RECENT STRESS

is greatly benefited and the need for grading is significantly reduced. Demand feeders will work with most species of fish with a little innovative and creative modification of the equipment used. Trout as small as 250 to the pound rapidly adapt to demand feeders.

The demand feeder concept has entered the electronic age. Scientists at the Fish and Wildlife Laboratory in Wellsboro, Pennsylvania have developed electronic mechanisms whereby fish can trigger feed delivery by swimming through a beam of light. The application of this concept is only limited by our creativity. What better way is there to meet the physiological requirements of the fish than to let the fish do it themselves. Our job is to develop the system and equipment to be used.

Dr. William Lewis and associates at the Fisheries Research Laboratory at Southern Illinois University at Carbondale have recently reported on their success in rearing striped bass (Table 9). Several features stand out as contributing significantly to their success of 80% survival rate.

1. Fish are fed very frequently, 16 times per day.
2. It appears that the fish are consistently overfed so that all the fish have easy access to feed which avoids size stratification and cannibalism. Initial feeding rates are between 25% to 50% of body weight per day dropping to 15%, 10% and then 5% of body weight by the time fish are 40 to 45 days of age.
3. Fish at 30°C receive twice the amount of those fish reared at 20°C.
4. An adequate training period as the fish are shifted from one feed to another.
5. Even when the fish are 50 days old they continue to be fed at the rate of 10 times a day. Even at this high feeding frequency the feed is delivered slowly and gradually spread out over the tank to insure even distribution and utilization by the fish.

Summary

Little scientific literature exists on the subject of feeds and feeding of striped bass. Nevertheless,

Table 9. Example of a successful feeding program for striped bass.

Age of the fish[1] (days)	Food type[2]
5-11	live brine shrimp nauplii
12-17	pulverized flake and brine shrimp nauplii[3]
18-22	pulverized flake, pulverized starter and live brine shrimp nauplii
23-30	pulverized starter, starter and live brine shrimp nauplii
31-35	starter
36-40	starter and 2/64 salmon
41-45	2/64 salmon
46-50	2/64 salmon and 3/64 salmon
51-60	3/64 salmon
61-75	3/64 salmon and 4/64 salmon
76-	4/64 salmon

[1]When reared at 23 to 25°C.
[2]Combination diets mixed in equal portions.
[3]The first three dry feeds are fed at frequent intervals (12 to 16 times per day). During this period the fry are also receiving brine shrimp nauplii.

Lewis (1981)

successful rearing of striped bass fingerling has been accomplished using commercially available feeds.

Physiologically the striped bass could be defined as a warm water trout. Good performance has been obtained with trout and salmon feeds.

Optimum growth occurs at 24°C. Growth is rapid requiring frequent well distributed feedings of adequate quantities of feeds, the lack of which results in size variation and excessive cannibalism.

References

Bayless, J.D. 1972. Artificial propagation and hybridization of striped bass, *Morone saxatilis* (Walbaum). South Carolina Wildlife and Marine Resource Department.

Cowey, C.B. and J.R. Sargent. 1972. Fish nutrition. Academic Press, Vol. 10:383-492.

Cox, D.K. and C.C. Coulant. 1981. Growth dynamics of juvenile striped bass as functions of temperature and ration. Transactions of the American Fisheries Society 110:226-238.

Eldridge, M.B., J.S. Whipple, D. Eng, M.J. Bowers and B.M. Jarvis. 1981. Effects of food and rearing factors on laboratory reared striped bass larvae. Transactions of the American Fisheries Society 110:111-120.

Gorman, D.D. 1982. Histology of the striped bass. American Fisheries Society. Monograph Number 3:21-39.

Kerby, J.H., L.C. Woods and M.T. Huish. 1983. A review of methods, advances and probelms associated with culture of striped bass and its hybrids. In press.

Lewis, W.M. and R.C. Heidinger. 1981. Tank culture of striped bass. Illinois striped bass project, IDC F-26-R. Southern Illinois University, Carbondale.

Maynard, L.A. and J.K. Loosli. 1969. Animal nutrition, 6th Ed, McGraw Hill, New York, New York. USA.

Millikin, N.R. 1982. Effects of dietary protein concentration on growth, feed efficiency and body composition of age 0 striped bass. Transactions of the American Fisheries Society 111:373-378.

The National Research Council. 1981. Nutrient requirements of cold water fishes. National Academy of Sciences.

The National Research Council. 1977. Nutrient requirements of warm water fishes. National Academy of Sciences.

Rogers, B.A. and D.T. Westin. 1981. Laboratory studies on the effects of temperature and delayed initial feedings on the development of striped bass larvae. Transactions of the American Fishereis Society 110: 100-110.

Setzler, E.M., W.R. Boynton, K.V. Wood, H.H. Zion, L. Lubbers, N.K. Mountford, P. Frere, L. Tucker and J.A. Mihursky. 1980. Synopsis of biological data on striped bass, Monroe saxatilis (Walbaum). NOAA Technical Report, NMFS Circular 433. FAO Synopsis Number 121. US Department of Commerce.

Parasites and Diseases of Striped Bass

Andrew J. Mitchell

Infectious diseases pose few problems in the culture of striped bass, even though 70 parasitic and disease agents of the species are known. Many of these pathogens are present continually in culture situations, particularly ponds. For a disease outbreak to occur, stressful conditions on the fish and optimum environmental conditions for the disease organism are required. The key to preventing disease lies in maintaining a suitable environment for the fish, supplying a good diet, imposing no undue stress, and using prophylactic treatments when stress cannot be avoided.

The few disease problems that do occur vary greatly from one cultural facility to another. Most of the variation in disease among hatcheries is accounted for by differences in intensity of culture, phases of culture, water quality (dissolved oxygen, total hardness and alkalinity, temperature, and salinity), and management. These variations cause considerable difficulty when one is attempting to discern key disease problems of national significance.

In a telephone survey of 20 striped bass hatcheries in 13 states in 1982, only two infectious diseases were reported to be major problems at 5 or more of these units. In a survey by letter, 7 years earlier, involving 27 hatcheries, only three diseases were noted to be causing problems at 5 or more facilities.

Correct identification of the pathogens by field personnel is also a key problem. Confusion is apparent in identification of the grubs (white vs. yellow), the small trichodinids, the gill flukes, and the bacteria. One culturist even reported the presence of viral hemorrhagic septicemia -- a disease known to be strictly limited to salmonids outside the U.S. -- in his striped bass.

Adding to the difficulties of summarizing striped bass diseases and treatments is the fact that much information on diseases and treatments is in the form of personal communications from the field, and scientific determinations have often not been made. With all these problems in mind, one must look at a summary of striped bass diseases as a combination of best guesses coupled with known facts. In this summary an attempt is made to group disease agents according to their importance, to report the most effective treatments used against them, and to describe disease signs helpful in diagnosing fish pathogens.

INFECTIOUS ORGANISMS

Major Infectious Agents of Striped Bass

Among the more than 70 infectious agents of striped bass known, only 8 are considered important disease producers on a nationwide basis.

Columnaris disease, caused by *Flexibacter columnaris* and related bacteria, is probably the major freshwater disease threat. It can cause large losses in a short time among broodfish, fingerlings, and fry, and can be especially devastating during summer. It compromises the integrity of the mucus, skin, and gill epithelium and readily becomes systemic. Since it is an opportunistic pathogen, prophylactic measures should be taken whenever the fish are known to be stressed -- especially when they are handled.

Three other genera of bacteria, *Aeromonas*, *Pseudomonas*, and *Vibrio*, commonly cause a condition known as fin rot in brackish water. This condition is related to poor water quality, inadequate diet, overcrowding, temperature fluctuations, and handling. In fresh water, fin rot can be caused by *F. columnaris*, aeromonads, or pseudomonads. Unless treatment is given, losses will probably result.

Aeromonas, *Pseudomonas*, and *Vibrio* can also occur as systemic infections, causing mortalities in freshwater and brackish environments. Environmental stressors such as low dissolved oxygen and high ammonia and other toxins in the water may precipitate these bacterial outbreaks.

Another common condition known as red sore disease, so called because of the hemorrhagic areas that occur on the body of the affected fish, is caused by epistylids (stalked ciliated protozoans) and *Aeromonas* spp. The cost of treatment in ponds where it usually occurs is prohibitive, and mortalities can result, especially during warm weather, when this condition is most common.

The ubiquitous protozoan parasite *Trichodina* is commonly found on striped bass during fall and spring. When water conditions and temperatures favor the parasite, its population can multiply and overwhelm weakened or stressed fry and fingerlings. It does most of its damage to the gills and can cause light to heavy mortalities.

An infestation by the dinoflagellate *Amyloodinium*, (often referred to as *Oodinium*), is a major problem for striped bass culturists in brackish water. This agent infests the gills and body of the fish. Velvet disease, as it is sometimes called, is favored by crowded conditions in a pond or tank. It is difficult to eradicate and can kill most of the infected fish if they are not treated.

One of the most visible diseases, external fungus, caused by *Saprolegnia*, *Achlya*, and perhaps several other genera of fungi, often occurs after handling and in association with columnaris infections. Unless treatments are given infected fish may die. Prophylactic treatments of broodfish after they are handled are advised, to reduce the chance for development of external fungus.

Minor Infectious Agents of Striped Bass

Several organisms, many of which are commonly found on striped bass, occasionally cause epizootics. The striped bass culturist must be aware of these agents, but should expect few problems from them.

These include Lymphocystis virus, the bacterium *Pasteurella piscicida* and the ectoparasites *Ambiphrya* (*Scyphidia*), *Cryptobia*, *Chilodonella*, *Ichtyobodo* (*Costia*), *Ichthyophthirius* (Ich), small trichodinids, *Trichophrya*, *Gyrodactylus*, *Dactylogyrus*, and *Ergasilus*. Most, if not all, of these pathogens become established and cause disease only in debilitated or stressed fish.

All can cause mortality, except for the Lymphocystis virus, which renders fish unacceptable to the public by enlarging the epithelial cells and causing the fish to have a grotesque lumpy white appearance. These cells slough off and the fish returns to normal without any known aftereffect. *Pasteurella piscicida* is a systemic bacterial infection that caused massive mortality of striped bass in the Rappahannock River and a light mortality in the Chesapeake Bay. Mortalities in a culture situation have not been reported. *Gyrodactylus* is the only ectoparasite in this group that exclusively infests the body. Ich, *Ichtyobodo*, and *Chilodonella* can (like *Gyrodactylus*), cause mortality by disrupting the integrity of the skin and mucus but, like the rest of the ectoparasites mentioned, they also cause mortality by damaging or blocking the respiratory surface of the gills. The culturist can expect to see light loads of some of these parasites but need be wary only of Ich, *Ichtyobodo*, and *Chilodonella* at low levels.

Infectious Agents that Cause Problems in Special Circumstances

The yellow grub, *Clinostomum complanatum*, found mainly in the muscle tissue and under the skin, and the white grub, *Posthodiplostomum minimum*, usually in and around the organs of the visceral cavity, become problems only when the grubs' three hosts are present and extremely high infection levels are reached. The white grub is known to have caused mortality in striped bass only once, but the yellow grub poses a more persistent problem, although it occurs at only a few facilities. Along with striped bass, usually the heron and snails of the genus *Heliosoma* serve as hosts for the yellow grub, and the heron and snails of the genus *Physa* for the white grub. Control can be achieved only by elimination of one of the hosts.

An unusual nematode infection, caused by *Goezia* sp., found in the walls of the intestine, was responsible for mortalities in striped bass stocked from a Florida fish hatchery. It was believed to have been introduced when infected marine herring were ground and fed to striped bass. In some lakes where the infected bass were stocked the worm evidently completed its life cycle, because mortalities occurred the following year in striped bass and *Goezia* was also recovered from another fish species, *Tilapia aurea*. If establishment of this infection in any lake or facility is to be avoided, uncooked marine fish should not be used as feed.

Branchiomyces, the cause of gill rot, is a systemic fungal infection of the gills that is a long-standing problem for fish culturists in Europe and Asia, but has been reported only once in striped bass in the United States. The spores of this fungus are believe to occur worldwide but a complex set of poorly understood environmental factors must be present before the disease develops. In Europe it is a summer disease favored by organically rich waters and temperatures above 20°C, and often is associated with the rearing of ducks. The disease, often has run its course by the time a diagnosis is made, and rarely recurs. Treatments are, at best, of questionable value but it is important to avoid excessive organic buildup.

Infectious Agents Not Associated With Disease Or That Have An Unknown Impact

Other organisms that have been found in striped bass and that either cause no known disease or that have an undetermined significance include 41 other parasites and two types of chlamydial agents (Table 1).

NONINFECTIOUS PROBLEMS

At least five noninfectious problems regularly occur in striped base culture. Two of these conditions involve dissolved gas levels in the water. Low dissolved oxygen is a common condition in densely stocked, organically rich ponds in summer. It can occur in any pond after a die-off of algae and is common in hot weather after several cloudy days. Nitrogen supersaturation, the main cause of gas bubble disease, often occurs in raceways. Water can be supersaturated with nitrogen if a pressure source (e.g., a deep well, high waterfall, restricted pipe with an air leak) is present in the water supply. Mortality can occur when supersaturation reaches 110%. Striped bass fry are reported to die at low levels (103% nitrogen) of supersaturation.

Soft waters with a total hardness of less than 20 ppm can cause striped bass to go into shock (as shown by such signs as flared gills, stiffening, erratic swimming, and jumping) when excited. The problem is more severe and mortalities can occur if the hardness is less than 5 ppm.

Problems with pH fluctuations occur in poorly buffered waters (low total alkalinity). In highly buffered water with dense blooms, the pH level can rise above 10, stressing and killing fry. However, the mortality associated with pH problems is seldom high.

Nutritional deficiencies that occasionally occur are often the result of either a poor batch of feed or inadequate zooplankton for the fry in the pond. Zooplankton can be adversely affected by the use of Masoten (Dylox) or Baytex for predacious insects. Unless the deficiencies are corrected, the loss of fish can be complete.

A number of other anomalies, including the improper development of the swim bladder in fry, spinal curvature, pugheadedness, and blindness, have also been reported.

TREATMENT AND CONTROL

The following account of the most frequently used controls found in the literature and reported from the field is not intended as a set of recommendations. Because treatment concentrations and lengths vary with the size and age of fish, water quality, temperature, and a host of other factors, bioassays on small lots of fish are advisable when one is using unfamiliar chemicals.

Table 1 lists most of the known infectious agents of striped bass and gives treatments that have been reported successful. Treatment rates, times and comments follow here.

The use of drugs or chemicals must be in accordance with current laws and regulations. Mention of product names does not imply endorsement by the U.S. Fish and Wildlife Service. The user should always read labels and follow all precautions and warnings.

Pond Treatments

The usefulness of copper sulfate ($CuSO_4$) is affected by the total alkalinity or the buffering capacity of the water. At an alkalinity of 200-300 ppm, up to 5.4 lb of $CuSO_4$/A-ft can be applied safely; at alkalinities below 40 ppm, $CuSO_4$ should not be used. As a general rule of thumb, for every 100 ppm alkalinity, 1 ppm $CuSO_4$ can be used. Most culturists use less than the calculated amount, but this practice may necessitate retreatment. At alkalinities above 300 ppm, the effectiveness of $CuSO_4$ may be limited. Some report it to be ineffective at temperatures below about 45° F. The powder or snow forms are recommended because the larger crystals dissolve too slowly. Low oxygen levels may result indirectly from its use in summer, because $CuSO_4$ kills algae and decomposing organic matter uses oxygen. To avoid shortcomings and dangers in its use, it is important that one first determine the total alkalinity of the pond water. One can purchase small, inexpensive, easily operated water analysis kits to determine the total alkalinity of the pond water.

For Branchiomycosis outbreaks in Europe and Asia, repeated $CuSO_4$ treatments throughout the course of the disease have been reported to be helpful.

The addition of lime in the form of agricultural limestone or calcium hydroxide (hydrated or slaked lime) to either the soil or water is an effective method of raising the total hardness and total alkalinity and avoiding soft water shock and pH fluctuations. About 30 to 50 pounds per surface acre of water or 1 to 2 tons per surface acre of bottom can be added to a pond. Higher levels of lime can be added to highly buffered water. About 200 pounds per surface acre reportedly gives good results against Branchiomycosis.

Masoten (Dylox) (80% active) is most effective as a parasiticide at water temperatures (°F) in the 60's and low 70's. When the water temperature is between 80 and 85°F, Masoten must be used early in the morning because it breaks down rapidly as the water warms. Masoten should not be used at temperatures above 85°F because it degrades so rapidly that it has no effect. Although rates of 0.25 ppm (0.85 lb/A-ft) have been recommended for controlling external flukes, 0.5 ppm (1.7 lb/A-ft) has been reported to be more consistently effective. A second treatment may be needed in about 3 days. *Ergasilus* should die after a 0.25-ppm treatment. For *Lernaea*, the anchor parasite, Masoten should be used at the lower rate once a week -- every 7th day -- for 4 weeks. The use of Masoten with fry should be avoided because the chemical kills zooplankton, and thus may cause starvation of the fry.

The effectiveness of potassium permanganate ($KMnO_4$) depends on the organic content of the pond. Eight pounds per acre-foot can kill certain fish species in "clear" waters, whereas 30 lb/A-ft may not even begin to treat the problem, let alone kill fish, in highly organic ponds. To obtain an effective treatment, one must use just enough chemical to maintain a light reddish-purple color in the water for about 12 hours. This can be done by applying 2 ppm (5.4 lbs/A-ft) at a time, until the desired color persists. If the color rapidly fades or turns brown soon after application, another 5.4 lbs/A-ft can be added. However, if it slowly fades after about 1 hour or more, only about 2-3 lb/A-ft should be added. In organically rich water, effective treatments may be cost prohibitive. The powder form is most easily dispersed. Because $KMnO_4$ kills algae, dissolved oxygen problems may occur when it is used in summer.

Tank or Bath Treatments

Acriflavin (trypaflavine) is an antibacterial that has been reported effective at 5-10 ppm (19-38 mg/gal; 0.67-1.3 oz/1000 gal, 5-10 oz/1000 ft^3). It is used at this rate for 1 hour or longer. Fry may succumb to extended treatment. As a prophylactic it has been used at 2 ppm.

Combiotic (penicillin plus streptomycin) used on external bacterial infections at 10 to 15 ppm (1.3 to 2 oz/1000 gal) for up to 24 hr can be repeated daily for several days. Some use it at 25 ppm for 2 hr on fingerlings.

Copper sulfate can be used as a bath treatment for protozoons and external bacteria at the rate given for pond treatments. It should not be used on fry. Treatments should last 1 hr or more.

Diquat (25% active) is reported as effective against *Flexibacter columnaris* and other external bacteria. It is effective in preventing some fungal infections on fish eggs and also has been reported to be effective on fish fungal infections. Application of 1 to 2 ppm (4 to 8 oz/1000 ft^3) have been used against bacterial problems. More Diquat may be needed if the organic load in the water is heavy and the chemical may be toxic to fry if used for extended periods.

Furacin (Nitrofurazone) is a broad-spectrum antibacterial that is effective against active *F. columnaris* infections, and many striped bass culturists have reported it to be effective against protozoan parasites and fungal infections. If these reports are accurate, the efficacy might be due to some of the additives that make up 90-95% of the formulated product; e.g., potassium dichromate, a strong oxidizer like $KMnO_4$, is found in one of the 9.2% active formulations. Reports of use were given at 100 ppm of the whole formulated product. Since the 4.59% active formulation was used, this would equal about 4.6 ppm. Several different formulations, 4.59, 9.2, 9.4, and 100% active ingredient, are now being used; therefore a suggested figure of about 5 ppm active ingredient is given to maintain consistent treatments. To yield 5 ppm when the 4.59% formulation is used, one applies 0.9 lb/1000 gal or 6.8 lb/1000 ft^3. When one is using the 9.2% or 9.4 % formulations, the rates should be reduced by one-half. Caution is required when the chemical is used on fry. The medication should be in the water at least 1 hr and usually can be left there indefinitely. Some have reported an avoidance reaction in fingerlings when Furacin is used in galvanized containers.

Furanace (Nifurpirinol) is an excellent treatment for columnaris disease. For striped bass, a treatment of 1 ppm active ingredient for 8 hr is suggested.

Formalin (37% commercial formaldehyde) is an effective tank treatment for many ectoparasites; however, it should not be used on fry and may kill fingerlings if they are stressed or debilitated by other disease problems. It has been used at 150 ppm (1.2 pt/1000 gal) for up to 1 hr. Before such a treatment is used, a rapid flush system should be set up in case fish show signs of stress early and, as a further precaution, fish should not be fed for at least 24 hr before treatment begins. Other factors to take into consideration are the need for abundant aeration during treatment, the fact that stress is more likely to develop when fish are crowded during treatment, and the fact that in soft acid water the treatment becomes more toxic. If the culturist is unsure, it is suggested that the treatment level be lowered to 50-100 ppm for 30 min to 1 hr.

Oxytetracycline (Terramycin) is an effective water bath for external bacterial infections and is often used at 10-20 ppm. When one is using TM_{50} (11% active), 12.2 to 24.3 oz/1000 gal gives a 10-20 ppm active level. Treatments should remain in the water for at least 1 hr.

Potassium permanganate ($KMnO_4$), a strong oxidizer, readily kills bacteria on the surface of fish. In clear water it is usually used at 2-3 ppm (0.27-0.39 oz/1000 gal or 2-3 oz/1000 ft^3) on striped bass. This treatment can be left indefinitely for striped bass other than fry. Fry should be removed at the first sign of stress or after 1 hr.

Quinacrine hydrochloride is sometimes used on velvet disease. Fish are first put in a freshwater (or low-salt) bath until they are stressed and then treated for 2 hr with 15-20 ppm active ingredient. One or two retreatments are advised for the next day or two. Only partial elimination of the parasite is accomplished, but further pathological effects are usually eliminated. Copper sulfate (0.5 ppm) has been substituted for quinacrine in this treatment, with about the same results.

Salt (NaCl) at 0.3% as a flush treatment has been reported to be effective on myxobacteria and protozoans. Others report the use of salt at up to 1% (83 lb/1000 gal). Salt at 1% is commonly used in combination with Furacin and Combiotic (rates discussed eariler) and is considered an excellent prophylactic or disease treatment. Salt can probably be used in combination with any of the above tank treatments. Salt treatments up to 1% can be left for at least 24 hr. Higher levels (3%) are reported toxic, especially to fry, in 24 hr. For epistylid infections a 1.5% concentration for 3 hr is required to effect complete control.

Most of the tank treatments make good hauling

prophylactics. For long hauls the concentrations may be reduced. A satisfactory oxygen level must be maintained in all tank treatments and the fish must not be crowded.

Medicated Feed

For active systemic bacterial infections medicated feeds are the only economical and practical pond treatment method. Three have been reported to be useful against bacterial diseases of striped bass.

Oxytetracycline (Terramycin) is the most widely used. It is fed at 3-4 g (active ingredient)/100 lb of fish for 10-14 days (3.5-4.7 oz active ingredient/100 lb feed or 2 to 2.7 lb TM_{50} [11% active]/100 lbs of feed for 10-14 days when fish are fed at 3% body weight). If fish are feeding at less than 3% body weight, more medication must be put into the feed to ensure that the fish receive the recommended amount.

Sulfamerazine is fed at 250 mg/kg (11.3 g/100 lb) of body weight per day for 14 days in the feed (13.3 oz/100 lb of feed fed at 3% body weight).

Chloramphenicol (Chloromycetin) is reported as the treatment of choice against *Pasteurella* disease when mixed with food at 50 mg/kg (2.3 g/100 lb) of body weight and fed for 5 to 10 days.

The above drugs are usually put on feed by "top coating" with vegetable oil or animal fat. Some can be purchased premixed in a fish feed.

Resistance of bacterial pathogens to fishery chemotherapeutants has been long observed. Estimates of resistance levels are as high as 80% in some bacterial species. Sensitivity testing should be done so that the effectiveness of an antibacterial can be determined before its application. This effort could produce monetary savings and prevent the further buildup of resistance by bacteria in the environment to a fishery chemical.

Di-N-butyl tin oxide is an antihelmintic that has been successfully used against some tapeworms and acanthocephalans of fishes. It is reported to be useful against *Pomphorhynchus rocci*, an acanthocephalan of unknown significance to striped bass culture, at a rate of 100 mg/lb of fish weight fed for 5 days.

Prophylactic Treatments

Prophylactic treatments, especially against columnaris disease and external fungal infections, are advisable after handling, exposure to low dissolved oxygen, or other stresses. (Prophylactic treatments are given under pond and tank treatments for the appropriate pathogens.)

Snail and Bird Control

Snails cannot be controlled by using available chemicals if fish are in a pond. When the fish are removed, however, copper sulfate at about 10 ppm (27.2 lb/A-ft) kills most of the snails. Also, if the pond is drained down to the pot holes, 10 ppm available chlorine (_HTH_ - 1.9 oz/1000 gal; _household bleach_ - 1.6 pt/1000 gal) kills the snails. Complete draining, drying, and disking of the pond bottom is the best method if time permits. If pond bottoms remain damp, _quicklime_ (used with extreme caution by persons wearing protective clothing) or fresh _slaked_ or _hydrated lime_ applied heavily to damp areas or pot holes, kills most snails.

The species of birds that are known to harbor parasitic stages of fish pathogens are protected by state and federal laws. Therefore, one must check with state and federal wildlife authorities before attempting to remove birds. Permission to kill such birds is often granted. Also, frightening devices and other methods can be recommended by local fish and game personnel. Pond-side perches such as trees should be removed, to prevent excessive bird droppings, (which may contain eggs of parasitic grubs) from entering the ponds.

DESCRIPTION AND DIAGNOSIS

The small number of disease problems that may be encountered by personnel at an individual striped bass cultural facility makes it possible for them to be trained in identification and treatment of pathogens. The major obstacle for these personnel is the difficulty of the isolation techniques and drug sensitivity testing that must be done on systemic bacterial pathogens. If persons are not properly trained, a nearby competent diagnostician should be sought. The "shotgun" approach -- the use of medicated feeds without pathogen isolation and sensitivity testing -- is costly, often ineffective, and selects for further resistance of bacterial pathogens in the environment.

Signs are helpful in diagnosing fish pathogens, but they can often be misleading. One must use the disease sign only as an aid in determining the presence of a fish pathogen; e.g., necrotic gill tissue indicates the presence of a gill pathogen. In almost all cases, proper diagnosis can be made only if the pathogen is found accompanying the sign.

Systemic Bacteria

Systemic bacteria (*Aeromonas*, *Pseudomonas* and *Vibrio*) can cause a variety of disease signs, including hemorrhage in the fins, at the base of the fins, on the skin, in the muscle, around the vent, and in the area of the isthmus; ulcers and skin lesions; raised and loosened scales; exophthalmia (popeye); ascites (belly bloat); and fin fraying. Internal signs include hemorrhage in and around the visceral organs and intestinal tract, ascitic fluid in the body cavity, and fluid in the intestines.

Pasteurella infections in striped bass are not visible externally, but internally the bacterium produces extensive bacteremia, often white nodules appear in the viscera.

Aeromonas and *Pseudomonas* species cause a septicemic condition called red vent in striped bass. The abdomen of the bass becomes distended and the anal opening enlarges and is inflamed. A yellowish discharge is often observed.

Behavioral characteristic of striped bass with bacteremias include listlessness, loss of appetite, clustering, and floating with the head up.

Isolates of systemic bacterial pathogens can usually be made from the kidney, liver, spleen, or any lesions.

External Bacteria

The most common sign associated with external bacterial pathogens (*Flexibacter columnaris*, *Vibrio* spp., *Pseudomonas* spp., and *Aeromonas* spp.) is fin or tail rot. The process of fin rot starts with the erosion of the fin membrane, leaving the unsupported fin rays with a ragged or frayed appearance. As the bacterial damage progresses, the rays are also eroded, leaving little or no externally visible fin tissue. *Flexibacter columnaris* continues to erode the base of the fin and surrounding muscle, eventually exposing the spinal column supporting the caudal peduncle.

The presence of the columnaris pathogen is also associated with gill necrosis, lesions with little or no hemorrhage, dull gray patches on the body caused by mucus loss, scale loss, and erosion of the mouth. Often areas of infection take on a yellow-brown hue because of the color of the bacterial colonies.

Behavioral signs are similar to those for systemic bacteria.

Microscopic examination is necessary to determine the presence of external bacteria, and isolation and biochemical testing techniques are necessary for identification.

Ectoparasites

Microscopic ectoparasites can cause gills to become bright red, frayed, uneven, puffed, or swollen.

Some gill parasites such as *Ergasilus* appear as recognizable white spots and can be identified by the use of a hand lens. However, most can be definitively identified only with the aid of a microscope.

Ectoparasites on the skin and the fins usually do not cause visible disease signs. However, petechial hemorrhage has been observed in areas of greatest irritation when these parasites are present in large numbers. A few of those reported that are visible to the naked eye include Ich (white spot), *Lernaea* (anchor parasite), Epistylid colonies (a mucus-like gray mass often surrounded by small hemorrhagic areas), and *Argulus* (a disk-like fish louse up to ¼ inch in diameter). Often fish with ectoparasites are seen flashing, crowding the bank or inlet, or with flared opercular flaps.

For generic identification of most ectoparasites a number of useful keys are available. One of the best is *Parasites of North American Freshwater Fishes* (Hoffman 1967).

Internal Parasites

A number of parasites, including protozoans and helminths occur within the muscle or viscera of the fish. Intestinal tapeworms and acanthocephalans appear as small white ribbonlike or stalklike projections from the mucosa (the inner membrane) of the intestine. It is suspected that emaciation of striped bass in some populations is caused by intestinal worms.

Yellow grubs may be seen through the skin as yellow-white raised areas up to ¼ inch in diameter, or they may be deep in the musculature. White grubs appear as small white spots in and on the visceral organs and tissue. Accurate identification of these and other internal parasites can be made only with the aid of a microscope.

Fungi

External fungal infections are probably the most apparent infection of fish. Small to large cotton-like patches form on the skin, fins, scales, and gills. Suspended organic and inorganic material in the water can collect in the hyphae (branching filaments), causing the color of the fungus patch to take on that of the suspended material. For positive identification, microscopic examination is necessary.

Branchiomyces can be seen and identified only microscopically, within the capillaries of the gill filaments and lamellae. The capillary blockage caused by the hyphae produces necrotic brown areas in the gills. After the infection has run its course, large portions of the gill tissue are often missing.

Table 1. Treatments for infectious and noninfectious agents of striped bass.

Agents	Treatment		
	Pond	Tank	Medicated feed
Major agents of importance			
Bacteria			
Flexibacter columnaris *Aeromonas* spp. *Pseudomonas* spp. *Vibrio* spp.	$CuSO_4$ $KMnO_4$	Acriflavin Combiotic $CuSO_4$ Diquat Furacin Furanace Terramycin $KMnO_4$ Salt	Terramycin Sulfamerazine
Fungi			
External fungus	CuSO $KMnO_4$	$CuSO_4$ Diquat? Furacin? $KMnO_4$	-----
Parasites			
Protozoa			
Epistylids	-----	Salt	-----

Trichodina spp.	$CuSO_4$ $KMnO_4$	$CuSO_4$ Formalin Furacin? $KMnO_4$ Salt	-----
Amyloodinium (*Oodinium*)	-----	$CuSO_4$ Quinacrine hydrochloride	-----
Minor agents of importance			
Virus			
Lymphocystis	-----	-----	-----
Bacteria			
Pasteurella piscicida	-----	-----	Chloramphenicol
Parasites			
Protozoa			
Ambiphrya *Chilodonella* spp. *Cryptobia* (*Bodomonas*, *Colponema* - complex) *Ichtyobodo* (*Costia*) *Ichthyophthirius* (Ich) Trichodinids (small)	$CuSO_4$ $KMnO_4$	$CuSO_4$ Formalin Furacin? $KMnO_4$ Salt	-----

Table 1. Continued

Agents	Treatment		
	Pond	Tank	Medicated feed
Trichophrya	$CuSO_4$	$CuSO_4$	-----
Trematode - Monogenea			
Dactylogyrus spp. *Gyrodactylus* spp.	Masoten $KMnO_4$	Formalin $KMnO_4$	-----
Crustacean			
Ergasilus	Masoten	Formalin Salt	-----
Special agents of importance			
Parasites			
Trematode - Digenea			
Clinostomum complanatum (yellow grub) *Posthodiplostomum minimum* (white grub)	Snail and Bird control	-----	-----

Cestode			
Goezia	-----	-----	-----
Fungi			
Branchiomyces	$CuSO_4$ Lime	-----	-----
Agents of little or questionable significance			
Chlamydia	-----	-----	-----
Epitheliocystis - small type			
Epitheliocystis - large cyst			
Parasites			
Protozoans			
Apiosoma	$CuSO_4$ $KMnO_4$	$CuSO_4$ Formalin $KMnO_4$ Salt	-----
Myxosoma morone	-----	-----	-----
Kudoa cerebralis			
Nosema			

Table 1. Continued.

Agents	Treatment		
	Pond	Tank	Medicated feed
Trematodes - Mongenea (external flukes)			
Ancyrocephalinae *Cleidodiscus pricei* *Urocleidus hastatus* *Diplectanum collinsi*	Masoten $KMnO_4$	Formalin $KMnO_4$	-----
Aristocleidus hastatus			
Trematodes - Digenea (internal Flukes)			
Lepocreadium setiferoides *L. areolatum* *L. californionum* *Neochasmus soyandoesi* *Stephanostomum tenue* Opecoelids - immature *Uvulifer* sp. *Neascus* sp. *Diplostomulum flexicaudum*	Snail and bird control where applicable	-----	-----

Diplostomulum sp.			
Ascocotylid-like species			
Cestodes (tapeworms)			
Proteocephalid larvae	-----	-----	Di-N-butyl. tin oxide for intestinal forms only
Scolex pleuronectis			
Trypanorhynchid plecrocercoid			
Rhynchothrium specinsium			
R. bulbifer			
Lacistomynchus sp.			
Nematoda (roundworms)			
Philometra rubia	-----	-----	-----
Cucullanus sp.			
Spinitectus sp.			
Contracaecum brachyurum			
Spiroxys			
Acanthocephala (thornyhead worm)			
Leptorhynchoides thecatum	-----	-----	Di-N-butyl tin oxide
Pomphorhynchus rocci			
Annelida - Hirudinea (leech)			
Myzobdella lugubris	Masoten	Salt	-----

Table 1. Continued.

Agents	Treatment		
	Pond	Tank	Medicated feed
Crustacea			
Achtheres lacae	Masoten	-----	-----
Argulus bicolor			
Caligus sp.	?		
Lironeca ovalis	?		
Aegathoa cf. oculata	?		
Lernaea	Masoten	Salt	-----
Mollusca (larval clam)			
Glochidia	-----	-----	-----
Vertebrata			
Rissola marginatta (cusk eel)	-----	-----	-----
Non infectious agents			
Low dissolved oxygen	Aeration	Aeration	-----
Supersaturation	Agitation	Agitation	-----
Soft water	Lime	-----	-----
Problems with pH	Lime	-----	-----

REFERENCES

Allen, K.O. 1974. Notes on the culture of striped bass in tanks and small raceways. Prog. Fish-Cult. 36(1):60-61

Bauer, O.N., V.A. Musselius, and Y.A. Strelkov. 1969. Diseases of pond fishes. Izdatel'stvo "Kolos," Moscow, 220 pp. (English translation TT72-50070, National Technical Information Service, U.S. Department of Commerce, Springfield, Va. 22151.)

Bonn, E.W., W.M. Bailey, J.D. Bayless, K.E. Erickson, and R.E. Stevens (editors). 1976. Guidelines for striped bass culture. American Fisheries Society, Striped Bass Committee of the Southern Division, 103 p.

Cordonnier, L.M., and H.L. Ward. 1967. *Pomphorhynchus rocci* sp. n. (Acanthocephala) from the rock bass, *Roccus saxatilis*. J. Parasitol. 53(6):1295-1297.

Davis, H.S. 1953. Culture and disease of game fishes. Univ. Calif. Press, Berkeley. 332 p.

Denoncourt, R.F., and W.H. Bason. 1970. Blind striped bass, *Roccus saxatilis* (Walbaum), from the Delaware River Estuary. Chesapeake Sci. 11(2):132-134.

Edwards, S.R., and F.M. Nahhas. 1968. Some endoparasites of fishes from the Sacramento-San Joaquin Delta, California. Calif. Fish Game 54(4):247-256.

Gaines, J.L., Jr. and W.A. Rogers. 1972. Fish mortalities associated with *Goezia* sp. (Nematoda: Ascaroidea) in Central Florida. Proc. Annu. Conf. Southeast Assoc. Game Fish Comm. 25:496-497.

Giudice, J.J., D.L. Gray, and J.M. Martin. 1981. Manual of bait fish culture in the south.

University of Arkansas Cooperative Extension Service and the U.S. Fish and Wildlife Service. 50 pp.

Hamer, P.E. 1966. The occurrence of striped cusk-eel, Rissola marginata, in the body cavity of striped bass, Roccus saxatilis. Chesapeake Sci. 7(4):214-215.

Hawk, J.P. 1976. A survey of the disease of striped bass, Morone saxatilis and pompano, Trachinotus carolinus cultured in earthen ponds. Proceedings of the World Maricult. Soc. 7:495-506.

Hensley, G.H., and F.M. Nanhhas. 1975. Parasites of fishes from the Sacramento-San Joaquin Delta, California. Calif. Fish Game. 61(4):201-208.

Hoffman, G.L. 1967. Parasites of North American freshwater fishes. Univ. of Calif. Press, Berkley. 486 p.

Hoffman, G.L., and J. A. Hutcheson. 1970. Unusual pathogenicity of a common metacercaria of fish. J. Wildl. Dis. 6(1):109.

Hughes, J.S. 1968. Toxicity of some chemicals to striped bass (Roccus saxatilis). Proc. Annu. Conf. Southeast Assoc. Game Fish Comm. p.230-234.

Hughes, J.S. 1973. Acute toxicity of thirty chemicals to striped bass, Morone saxatilis. Louisiana Fed. Aid Fish Rest. Project F-15-6. 15 p.

Johnson, C.A., and R. Harkema. 1970. A preliminary report on the life history of Pomphorhynchus rocci, an acanthocephalan parasite of striped bass. J. Elisha Mitchell Sci. Soc. 86(4):184-185.

Krantz, G.E. 1970. Lymphocystis in striped bass, Roccus saxatilis, in Chesapeake Bay. Chesapeake Sci. 11(2):137-139.

Linton, E. 1924. Notes on cestode parasites of sharks and skates. Proc. U.S. Natr. Mus. 64(art. 21):1-114.

Mahoney, J.B., F.H. Midlige, and D.G. Deuel. 1973. A fin rot disease of marine and euryhaline fishes in the New York Bight. Trans. Am. Fish. Soc. 102-(3):596-605.

Meyer, F.P., and J.A. Robinson. 1973. Branchiomycosis: A new fungal disease of North American fishes. Prog. Fish-Cult. 35(2):74-77.

Mitchell, A.J., and G.L. Hoffman. 1981. Preparation of live channel catfish for shipment to states requiring a health permit. Aquacult. Mag. 7(6):28-29.

Overstreet, R.M. 1971. *Noechasmus sogandaresi* n. sp. (Trematoda:Crytogonimidae) from striped bass in Mississippi. Trans. Am. Microscop. Soc. 90(1):87-89.

Paperna, I., and D.E. Zwerner. 1974. *Kudoa cerebralis* sp. n. (Myxosporidea, Chloromyxidae) from the striped bass, *Morone saxatilis* (Walbaum). J. Protozool. 21(1):15-19.

Paperna, I., and D.E. Zwerner. 1976. Parasites and diseases of striped bass, *Morone saxatilis* (Walbaum), from the lower Chesapeake Bay. J. Fish Biol. 9:267-281.

Plumb, J.A. (editor). 1979. Principal diseases of farm raised catfish. Southern Cooperative Series No. 225. School of Agriculture, Auburn University, Auburn, AL. 92 p.

Regan, D.M., T.L. Wellborn, Jr., and R.G. Bowker. 1968. Striped bass, *Roccus saxatilis* (Walbaum), development of essential requirements for production. U.S. Department of Interior, Fish and Wildlife. Serv., Division of Fish Hatcheries, Atlanta, Ga. 133 p.

Sindermann, C.J. 1970. Principal diseases of marine fish and shellfish. Academic Press, New York. 369 p.

Snieszko, S.F., G.L. Bullock, E. Hollis, and J.G. Boone. 1964. Pasteurella sp. from an epizootic of white perch (Roccus americanus) in Chesapeake Bay tidewater areas. J. Bacteriol. 88(6):1813-1814.

Stickney, R.R. 1979. Principles of warmwater aquaculture. John Wiley and sons, N.Y. 375 p.

Wellborn, T.L., Jr. 1969. The toxicity of nine therapeutic and herbicidal compounds to striped bass. Prog. Fish-Cult. 31(1):27-32.

Wellborn, T.L., Jr. 1971. Toxicity of some compounds to striped bass fingerlings. Prog. Fish-Cult. 33(1):32-36.

Wolke, R.E., D.S. Wyand, and W.H. Khairallah. 1970. A light and electron microscopic study of epitheliocystis disease in the gills of Connecticut striped bass (Morone eaxatilis) and white perch (Morone americanus). J. Comp. Pathol. 80(4):559-563.

Cryopreservation of Striped Bass Spermatozoa: Problems and Progress

Jerome H. Kerby

Striped bass (*Morone saxatilis*) have represented an important commercial and recreational resource in marine and estuarine waters of the United States since colonial times (Jordan and Everman 1902). The first hatchery for the species was established in 1884 on the Roanoke River in North Carolina (Worth 1884). Development of techniques for hormone-induced spawning by Stevens and his co-workers (Stevens et al. 1965; Stevens 1966, 1967) at a newly established South Carolina hatchery, and subsequent widespread successful use of the striped bass and its hybrids (principally striped bass X white bass, *M. chrysops*) for management and recreational purposes in inland waters has resulted in the establishment of a number of hatcheries in the southern United States.

Striped bass brood stock are usually collected on or near the spawning grounds. However, hatcheries sometimes have difficulty in obtaining an adequate supply of ripe males (Bayless 1972; Bonn et al. 1976; Texas Instruments 1977). These difficulties usually occur late in the spawning season, when ripe females may be abundant, but males are scarce. Further, because peak spawning of white bass typically occurs earlier than that of striped bass, the collection of male white bass for use in production of hybrids is sometimes difficult.

These difficulties stimulated my efforts to cryopreserve striped bass spermatozoa. Although the efforts were successful, improvements are needed to realize the full potential of the technique.

EXPERIMENTAL OBJECTIVES

Initial experiments were designed to identify suitable extenders and cryoprotectants for striped bass by testing those previously demonstrated to be effective for other fish species. Various techniques and freezing rates were also used in an effort to obtain the best possible results. Later experiments concentrated on developing methods for cryopreserving the semen in volumes suitable for use in production hatcheries and for reducing the variability of results between experiments. Survival and growth of fish produced with cryopreserved spermatozoa were also compared with survival and growth of fish produced with fresh semen.

EXPERIMENTAL FEASIBILITY

Experimental results demonstrated that striped bass sperm can be successfully cryopreserved by using several different extenders combined with dimethylsulfoxide (DMSO) as the cryoprotectant. Best overall results were obtained with extender OH-189, developed by Ott (1975), combined with 5.0% DMSO and mixed in a 1:4 sperm:medium (volume:volume) ratio. The highest fertilization percentage obtained with cryopreserved sperm was 87.7. Other extenders that also produced good results were OH-134, OH-235, and OH-275 (Kerby 1983). No fertilization was obtained when other cryoprotectants (glycerol, ethylene glycol or propylene glycol) were used. Mean freezing rates greater than 5°C/minute were more effective than slower rates (Kerby 1983).

Despite the fact that fertilization could be consistently obtained with cryopreserved sperm, and that fertilization capacity could be retained for up to two years, individual fertilization percentages were not consistent, and varied considerably from one trial to the next. Part of the inconsistency was due to variation in egg quality among females, and eggs from some females may be more "receptive" to the frozen-thawed sperm than those from others. Other, perhaps more important, variables that affected consistency included small variations in technique from one trial to the next (in both the freeze-thaw and the fertilization processes), the inability to obtain exactly repeatable freezing rates, and possible differences in semen quality. Overall fertilization percentages from cryopreserved sperm were seldom comparable with those of fresh sperm. Results of 48 individual tests with extender OH-189 showed the variability obtained at different freezing rates, with different sample lots, and with different females (Table 1).

PRODUCTION REQUIREMENTS

In later experiments I explored methods of cryopreserving the large volumes of semen that would be necessary in a production hatchery situation. Female striped bass normally produce from 1 to 3 liters of eggs, depending on the size of the fish. Fresh semen from at least 2 males is normally used, with a total volume usually ranging from 10 to 30 ml per batch of eggs. Since cryopreserved sperm cannot yet be expected to have a fertilization capacity equivalent to fresh sperm, larger volumes of semen are required. Inasmuch as the cryopreserved sperm is already diluted 1:4 with the extending medium, considerable quantities of material must be effectively preserved and stored.

Several types of freeze-storage containers were examined. Because generally satisfactory results were obtained in the initial experiments with a 2-ml polypropylene A/S NUNC screw-capped vial from Union Carbide Incorporated[1], I tested a 5-ml vial of the same diameter. This also provided satisfactory results, but volumes required, coupled with the logistics of thawing sufficient numbers of vials, rendered them unsuitable for hatchery use. A similar vial with the same diameter and about 300 mm long would contain about 20 ml of extended semen and might prove satisfactory. However, the manufacturer does not now make such a vial. Substitutes, consisting of rigid polypropylene and polyethylene tubing with similar

[1] Use of brand names does not imply government endorsement.

Table 1. Percent fertilization (mean ± SE; range in parentheses) of striped bass ova with sperm frozen and stored in liquid nitrogen (-196°C)[1].

Sample Lot Number	Mean Freezing Rate[2] (°C/min)	Number of Trials	Range in Time Frozen (Days)	Treatment of Sperm[3]: Cryopreserved[4]		Treatment of Sperm[3]: Fresh Control	
35	11.1	8	0.6-701	11.9 ± 4.3	(0-37)	64.4 ± 6.3	(33-87)
39	2.4	4	0.6-321	11.3 ± 4.4	(3-22)	68.5 ± 11.9	(33-86)
41	3.5	5	1.2-696	19.4 ± 7.5	(12-55)	64.9 ± 3.6	(56-77)
45	7.6	5	1.1-696	29.2 ± 7.8	(12-55)	55.4 ± 11.4	(29-92)
49	11.4	5	1.7-696	21.9 ± 7.4	(0-39)	61.9 ± 12.9	(28-89)
53	17.0	4	0.2-694	8.8 ± 2.2	(3-13)	56.9 ± 9.4	(29-72)
61	7.4	3	0.8-002	49.6 ± 19.6	(22-88)	89.2 ± 4.1	(82-96)
65	13.0	6	0.3-373	46.2 ± 3.9	(36-58)	61.3 ± 11.7	(11-89)
69	8.4	6	1.2-734	26.2 ± 7.3	(9-57)	72.8 ± 8.6	(36-92)
73	8.0	2	358	9.7 ± 4.8	(5-15)	49.6 ± 7.8	(42-57)

[1]Extender was OH-189. Cryoprotectant was 5.0% dimethylsulfoxide.

[2]Mean freezing rate (°C/minute) was calculated from the extremes of +10 to -40°C.

[3]Sperm:medium ratio was 1:4 (volume:volume). Each sample lot of a single male's semen frozen under the same conditions.

[4]Time frozen shows range of times that individual samples were frozen before fertilization trials.

diameters and cut to approximate lengths did not appear to provide results (% fertilization) equivalent to those obtained with the NUNC vials. Heat transfer from these containers may differ because their molecular construction is somewhat different and the walls are thicker. Freezing blood bags (plastic bags in which human blood is frozen) were also tested, but did not appreciably improve results. They also presented additional logistical problems from the standpoint of freezing, thawing, and storage. Only two to four bags can be frozen at one time in a freezing chamber, and large, expensive storage facilities would be needed because of the configuration of the bags. Configuration also resulted in thawing difficulties. Finally, cost for routine use would be prohibitive.

To date, fertilization percentages obtained by using large volumes of cryopreserved semen in a production situation have ranged as high as 55% and several million larvae have been produced during the last 4 years. However, consistency between trials has been poor, and often the fertilization percentage was very low. Improvements are required for the technique to be routine in production situations.

Further experiments, including the use of thin-walled stainless steel tubes and examination of a technique for freezing directly on solid CO_2 (-79°C) are contemplated. Electron microscopy will also be used in an effort to better determine the effects of various freezing and thawing rates and extenders on sperm morphology.

PRODUCTION OF FISH

Although several investigators (Blaxter 1953; Mounib et al. 1968; Graybill and Horton 1969; Stoss and Holtz 1981; and others) have reported hatching eggs fertilized with frozen-thawed sperm, only Moczarski (1977) demonstrated that larval common carp (*Cyprinus carpio*) produced with cryopreserved sperm had survival and growth rates comparable with those of fish produced with fresh sperm. Because such information is necessary to adequately determine the potential of cryopreservation techniques with regard to fish husbandry, Kerby et al. (1982) conducted a study to compare growth and survival of striped bass larvae hatched from eggs fertilized with cryopreserved sperm with those of larvae produced with fresh sperm. In two experiments, they stocked six 0.4-hectare and six 0.2-hectare ponds at an estimated rate of 250,000 larvae per hectare. More total fingerlings were harvested from ponds stocked with larvae produced with fresh sperm, but differences in the means were not significant in either experiment. Results were similar for harvest weights. Mean lengths and weights of fingerlings produced with cryopreserved sperm were significantly greater than those for fish produced with fresh sperm, but these differences were probably a function of relative density and food availability rather than of the treatments. Very few of the fish from either treatment were abnormal; most appeared healthy.

OVERVIEW

In my view, the potential for using cryopreservation techniques for striped bass and hybrid husbandry is tremendous, providing the problem of obtaining a con sistent product in sufficient quantities is solved. Methods are now sufficiently advanced that they can be used successfully for experimental breeding and hybridization programs. Available extenders appear to be capable of providing satisfactory results for striped bass, but work directed toward determining satisfactory extenders for other *Morone* species is also needed.

It is probably not necessary for fertilization results to be equal to those obtained with fresh sperm, and each hatchery manager can decide what levels would be required in a given situation. Average percentages of 30 to 50% might be adequate in some situations, whereas higher percentages would be required elsewhere.

The growth and survival experiments leave little doubt that the fish produced from frozen sperm are normal and healthy. Thus, there should be no hesitancy in using the procedures.

Support for the research described was provided in part by a faculty research grant from North Carolina State University, and in part by contracts 14-16-0008-2146 and 14-16-0009-132 from the United States Fish and Wildlife Service. Additional support and assistance were provided by the South Carolina Wildlife and Marine Resources Department. I thank Melvin T. Huish for reviewing the manuscript and Dorothy Wright for typing it. I also thank Jeff Hinshaw for presenting it at the Annapolis conference when personal reasons prevented me from attending.

ACKNOWLEDGEMENTS

The North Carolina Cooperative Fishery Research Unit is jointly sponsored by North Carolina State University, the North Carolina Wildlife Resources Commission, and the United States Fish and Wildlife Service, Department of the Interior.

REFERENCES

Bayless, J. D. 1972. Artificial propagation and hybridization of striped bass, *Morone saxatilis* (Walbaum). South Carolina Wildlife and Marine Resources Department. 135 pp.

Blaxter, J. H. S. 1953. Sperm storage and cross-fertilization of spring and autumn spawning herring. Nature (London) 172:1189-1190.

Bonn, E. W., W. M. Bailey, J. D. Bayless, K. E. Erickson, and R. E. Stevens, eds. 1976. Guidelines for striped bass culture. Striped Bass Committee of the Southern Division, American Fisheries Society. 103 pp.

Graybill, J. R., and H. F. Horton. 1969. Limited fertilization of steelhead trout eggs with cryopreserved sperm. Journal of the Fisheries Research Board of Canada 26:1400-1404.

Jordan, D. S., and B. W. Evermann. 1902. American Food and Game Fishes. Doubleday, Page and Company, New York. 572 pp.

Kerby, J. H. 1983. Cryogenic preservation of sperm from striped bass spermatozoa. Transactions of the American Fisheries Society 112:86-94.

Kerby, J. H., J. D. Bayless, and R. M. Harrell. 1982. Growth, survival, and harvest of striped bass produced with cryopreserved spermatozoa. Presented at the annual meeting of the American Fisheries Society, Hilton Head, South Carolina, September 25, 1982.

Moczarski, M. 1977. Deep freezing of carp *Cyprinus carpio* L. sperm. Bulletin de l'Academie Polonaise des Sciences Serie des Sciences Biologiques 25:187-190.

Mounib, M. S., P. C. Hwang, and D. R. Idler. 1968. Cryogenic preservation of Atlantic cod (*Gadus morhua*) sperm. Journal of the Fisheries Research Board of Canada 25:2623-2632.

Ott, A. G. 1975. Cryopreservation of Pacific salmon and steelhead trout sperm. Doctoral dissertation. Oregon State University. Corvallis, Oregon, U. S. A.

Stevens, R. E. 1966. Hormone-induced spawning of striped bass for reservoir stocking. Progressive Fish-Culturist 28:19-28.

Stevens, R. E. 1967. A final report on the use of hormones to ovulate striped bass, *Roccus saxatilis* (Walbaum). Proceedings of the Annual Conference of the Southeastern Association of Game and Fish Commissioners 18:525-538.

Stevens, R. E., O. D. May, Jr., and H. J. Logan. 1965. An interim report on the use of hormones to ovulate striped bass (*Roccus saxatilis*). Proceedings of the Annual Conference of the Southeastern Association of Game and Fish Commissioners 17:226-237.

Stoss, J., and W. Holtz. 1981. Cryopreservation of rainbow trout (*Salmo gairdneri*) sperm. I. Effect of thawing solution, sperm density and interval between thawing and insemination. Aquaculture 22:97-104.

Texas Instruments Incorporated. 1977. Feasibility of culturing and stocking Hudson River striped bass. An overview, 1973-1975. Texas Instruments Incorporated Ecological Services, Dallas, Texas.

Worth, S. G. 1884. Report upon the propagation of striped bass at Weldon, N. C., in the spring of 1884. Bulletin of the United States Fish Commission 4:225-230.

Overview of Legal Constraints on Aquaculture

Alex W. Wypyszinski

As sources of protein in the nation's diet, meat, poultry and dairy products rank much higher than fish. Nevertheless, fish landings in the United States have not increased at even the modest rate of fish consumption. Although it has been estimated that as much as 20%[1] of the world's harvestable supply of fish can be found on the continental shelf areas adjacent to our coastlines and elsewhere in the fishery conservation zone established by the Magnuson Fisheries Conservation and Management Act[2] (FCMA), most of the fish eaten by Americans is imported. In 1976, 63% of edible fishery products were imported, creating a balance of trade deficit of nearly two billion dollars.[3]

While the FCMA and other laws[4] were passed by Congress in an effort to increase the ability of the various segments of the U.S. fishing industry to capitalize on the potential for economic growth, the potential contribution of aquaculture seems to have been treated almost as an afterthought.

Aquaculture is based on an assumption that proper management of controlled systems, by permitting optimal use of input materials such as feed and energy, can provide greater yield than is possible in unmanaged natural systems.[5] Fish culture has been practiced for thousands of years, and today aquaculture facilities provide nearly 10% of the world catch.[6] Although it is a growing industry here, producing in excess of 90,000 metric tons of finfish and shellfish annually[7], the United States produces a negligible fraction of this percentage.

Conceptually, aquaculture is to fishing as agriculture is to hunting, but one authority estimates that U.S. aquaculture today is about where agriculture was three hundred years ago, "very low in technology, but very high in promise and opportunity."[8]

Like any commercial agricultural enterprise, commercial aquaculture is typically confronted by an array of federal, state and local legal requirements. Agriculture, however, has been able over the years to develop relatively coordinated policies and programs administered by comparatively few government agencies. Furthermore, agricultural programs for the most part enjoy a significant level of political support. Aquaculture has not been so fortunate. It is a relative newcomer that does not fit neatly into existing agricultural programs and as a result is regulated at each level of government by a number of agencies, bureaus, departments and offices; each with traditional "turf" and a tendency to preserve the status quo. Federal agencies are involved in programs ranging from financial and technical assistance to regulation of health and sanitation and environmental protection. State agencies are also involved in these areas as well as water use and fish and wild life management. Local governments usually regulate land use and construction. The majority of laws and regulations that specifically authorize permit or control aquaculture operations can be found at the state level, although generally all three levels of government are involved at each step of aquaculture development and operation. It is thus very easy to find regulatory gaps, inconsistant policy, duplication of effort and overlapping jurisdiction in a maze which is in its totality the principle regulatory barrier to the potential aquaculture entrepreneur.

It should also be noted that "constraints" such as health and sanitation regulations and many others are indisputably beneficial to the industry. Few people would suggest that this type of regulation should be made less stringent.

On reflection then, it would appear that the source of the problems created by the legal and regulatory regime affecting aquaculture has been the absence of direction and coordination. Despite a number of well intended programs there has been lacking a coherent national policy.

The federal government has been involved in aquaculture programs for at least a hundred years[9], but it was with passage of the National Aquaculture Act of 1980[10] that Congress finally attempted to establish a national policy to promote the development of aquaculture.

THE NATIONAL AQUACULTURE ACT OF 1980

An examination of the legislative history[11] of the Act indicated that Congress recognized the potential of aquaculture in the U.S. as well as the problems which affect the industry - including economic, production and legal factors which actually inhibit the development of aquaculture as a commercial enterprise. Legal and regulatory constraints were perceived as among the most constraining influences hindering such development.

The Act is notable in that it does not establish another licensing and regulatory framework, but rather establishes an interagency aquaculture coordinating group charged with an advisory role which includes the collection and dissemination of information as well as coordination of all federal activities affecting aquaculture. The Secretaries of Agriculture, Commerce and Interior are required to establish a National Aquaculture Development Plan[12] which includes " . . . programs to analyze, and formulate proposed resolutions of the legal or regulatory constraints that may affect aquaculture."[13] Furthermore, the Secretaries are required to make a continuing assessment of " . . . the economic, physical, legal, institutional and social constraints that inhibit the development of aquaculture in the U.S."[14]

Congress also ordered a study to be conducted of State and Federal regulatory restrictions to aquaculture development in the U.S.[15] This study was to include a literature search and a descriptor list identifying the parameters of the issue; a list of relevant current and pending state and federal regulations; and, case studies of a number of existing aquaculture operations to determine the practical effect of regulatory restrictions.

This latter congressional mandate resulted in an eighteen month research and writing effort and a six volume (unpublished) report[16] that specifically identifies the enormous and complex body of state and federal laws and regulations which either directly or indirectly affect the development of commercial aquaculture operations in this country.

According to the report there are approximately 50 federal statutes (accompanied by implementing regulations) which have a direct impact on how, when, where, and with what the fish farmer does business and over 120 federal statutory programs identified as having a significant relationship to aquaculture.[17] Furthermore, although researchers examined the official codes of only 32 of the 50 states, over 1200 state laws with varying degrees of impact on aquaculture development were identified.[18]

The National Aquaculture Development Plan required under §2803 of the Act is now in draft form and has been submitted to the Departments of Agriculture, Commerce and Interior for approval, and, according to the Aquaculture Coordinator at the Department of Agriculture, could be released within the next few months (depending on OMB review and Congressional appropriations).[19]

These developments at the federal level since passage of the National Aquaculture Act are encouraging, but changes in the regulatory climate are not likely to occur overnight.

OVERVIEW OF REGULATORY CONSTRAINTS

Despite the progress and potential for change created with passage of the National Aquaculture Act, it is very unlikely that any radical changes in the legal and regulatory system will occur quickly. The potential aquaculturist should therefore be aware of the general categories of direct and indirect restrictions which he is likely to encounter. Many of these restrictions would be faced by any small businessman; others by anyone seeking establishment of a water oriented enterprise; and still others are directed particularly at the aquaculture industry. The net effect of these regulations is the creation of a need for considerable expenditures of time and money and the consequential creation of an atmosphere which discourages investment in the industry.

The following general categories of regulation should be considered by the potential aquaculture entrepreneur: land regulation; water regulation; pollution; fish and fisheries management; facility/hatchery management; and, processing operations. Additionally, the areas of commercial/financial regulation and labor policy have been identified as legal constraints, but these areas are best treated from an economic and marketing perspective. Obviously there is a good deal of overlap among these areas, and, particularly in the areas of land and water regulation, all three levels of government - federal, state and local - become involved in one degree or another.

Land Regulation

A. Zoning. In the eighteenth century, the elder Pitt declaimed that "the poorest man in his cottage could defy the King - storms may enter; the rain may enter - but the King of England cannot enter."[20] Following the American Revolution and well into the twentieth century, the only limitations placed on a person's use of private property were common law limitations such as those prohibiting uses construed to be public or private nuisances. The door was figuratively opened to the King in 1926, when the Supreme Court held that the sovereign power - the state - has the authority to restrict and regulate private property rights when such regulation is for the protection of the public health, safety, morals or general welfare.[21] Zoning is now the most widely employed form of land use control. Commonly delegated by the state to county or local authorities, a zoning ordinance is valid to control the use of private land unless it is found to be unreasonable, arbitrary, discriminatory or confiscatory. The state retains oversight authority over ordinances passed by county and municipal authorities, and the private landowner is further protected by the 5th Amendment to the federal constitution. That amendment prohibits the government (federal, state or local) from taking private land for public purposes without payment of just compensation. Such a taking is easy to find when the government wants the land to build a highway or a dam, but the question of takings is not so clear-cut where the government limits or prohibits a use of land.[22]

Very often, the acquisition and use of land as an aquaculture facility is a new use; not contemplated by zoning authorities and not designated under guidelines of a local master plan. The aquaculture developer is then faced with obtaining a variance or an amendment to the zoning code - both time consuming and potentially costly endeavors. Development constraints may also occur where uncertainty persists as to whether aquaculture is an agricultural or an industrial use. Furthermore, the aquaculturist, like any other developer, must comply with permitting requirements of building codes and construction standards.

B. Coastal Zone Management. In 1972, after consideration of a national land use law, Congress passed the Coastal Zone Management Act (CZMA).[23] The CZMA established a system of federal grants as incentives for individual states to develop enforceable programs for land and water use planning in the coastal zone. The CZMA defines the coastal zone as " . . . the coastal waters and the adjacent shorelands strongly influenced by each other and in proximity to the shorelines of the several coastal states and includes . . . transitional and intertidal areas, salt marshes, wetlands and beaches."[24]

In order to obtain the federal grants, the state must establish an approved program which provides means to administer coastal zone land and water use regulations, control development, and provide for a system of conflict resolution among competing uses. Although participation in the federal program is voluntary, most states now have approved coastal zone management programs.[25] Since the CZMA encourages rational, mixed use in the coastal zone, it is not disadvantageous to the aquaculturist - water dependent uses are generally encouraged.

Most often, the coastal zone management programs are based on a "networking" of existing statutes. In New Jersey for example, the three principal regulatory laws by which the Department of Environmental Protection manages the coastal zone are the Waterfront Development Law[26], the Wetlands Act[27], and the Coastal Area Facility Review Act[28] (CAFRA).

Installation of a new dock, pier, bulkhead or mooring in a tidal water body will require a waterfront development permit. An application for such a permit must contain engineering drawings prepared by a licensed professional engineer[29] and a site plan and survey depicting existing and proposed structures on the site, property lines and mean high and mean low water lines.

A wetlands permit is required for regulated activities on coastal wetlands. Such activities include excavation of small boat mooring slips, maintenance or repair of bridges, roads or highways, and construction of piers, catwalks, docks, landings and observation decks. In addition, a wetlands permit (Type B) is required for the installation of utilities, excavation for boat channels and mooring basins, construction of impoundments and sea walls, water diversion, and the use of pesticides.

CAFRA authorizes the Department of Environmental Protection to regulate and approve the location, design and construction of major facilities in an area which includes the bulk of the state's coastal zone. Among the facilities regulated by CAFRA and requiring a permit are all food and food by-product facilities.[30] A CAFRA permit application (and a Type B wetlands permit application as well) must contain an Environmental Impact Statement (EIS). State regulations require the EIS for a CAFRA permit to include: an inventory of existing environmental conditions at the project site and in the surrounding region which shall describe air quality, water quality, water supply, hydrology, geology, soils, topography, vegetation, wildlife, aquatic organisms, ecology, demography, land use, aesthetics, history and archaeology; a project description which shall specify what is to be done and how it is to be done during construction and operation; a listing of all licenses, permits or other approvals as required by law and the status of each; an assessment of the probable impacts of the project; a listing of adverse environmental impacts which cannot be avoided; steps to be taken to minimize adverse environmental impacts during construction and operation, both at the project site and in the surrounding region; alternatives to all or any part of the project with reasons for their acceptability or unacceptability; and, a reference list of pertinent published information relating to the project, the project site, and surrounding region.[31]

Before becoming involved in the permitting process, the applicant should be aware that a valid tidelands instrument[32] must be obtained prior to submitting an application for any of the three previously mentioned permits.

Tidelands are those lands which are now or were formerly flowed by the tides - the area between mean high water and mean low water. On flat, coastal plain areas (such as New Jersey) the amount of acreage can be enormous. Without a valid tidelands instrument a person occupying tidelands has the legal status of a trespasser; the state is entitled to obtain the fair market value of such land and the fair market rental value for the period during which it was illegally occupied.

In order to obtain a tidelands grant, lease or license the applicant must submit a current survey, prepared by a licensed surveyor, showing the upland property, the boundaries of the tidelands areas aplied for, the location of the mean high water line, the depth of the waterway at mean low water, the names of adjoining property owners, a diagram of proposed or existing structures within the area; and, a certificate of title signed by an attorney or title company representative showing that the applicant owns the upland property or has the permission of the upland owner to apply for the conveyance.[33]

There has been some movement towards streamlining

the system. The Office of Cultural and Environmental Services coordinates the review of major development proposals likely to require more than one DEP-administered permit. The Office of Business Advocacy in the State Department of Commerce helps developers determine which state permits are needed. Pre-application conferences are encouraged and potential developers are advised at an early stage whether a proposal is likely to be approved or what modifications would enhance the likelihood of approval.

Water and Pollution Regulation

Although jurisdictional boundaries on land are easier to identify than water boundaries, it is not terribly difficult to clarify these jurisdictional lines. Inland waters - rivers, streams, lakes and groundwater normally are regulated by the states subject to reserved rights of navigation and water quality preservation vested in the federal government.[34] That ocean area within three miles of a coastal state is known as the territorial sea and is also within state jurisdiction.[35] Seaward of the territorial sea out to 200 miles off the coast is administered by the federal government.[36] The high seas are governed by international law; both customary and treaty law.

In identifying the water resources necessary to an aquaculture operation and assuring a sufficient supply, it is necessary to determine the ownership or control of that water. In most circumstances water is considered a public resource and as such competition for that resource potentially involves shipping, waste disposal, commercial and recreational fishing, and boating interests. Here again, statutory silence about water use for aquaculture could be viewed as a legal constraint.

Riparian rights in most states include rights of access to and from the water, "wharfing out" rights, and preference in development of adjacent submerged land.[37]

However, the law regarding issues of riparian and littoral water rights is complex. The aquaculturist must be aware of applicable state laws dealing with acquisition of such rights, and must insure that the aquaculture operation does not unduly infringe on the rights of other riparian owners.

Water pollution can be either a threat to stock or a by-product of aquaculture operations (or both). Furthermore, the potential fish farmer will find that siting options are limited; often due to illegal discharges and less than stringent enforcement of existing regulations. On the other hand, the aquaculturist will find that the discharge of effluents from ponds and raceways will normally require a permit.

Any discharge of a pollutant from a point source into U.S. waters is prohibited unless made pursuant to a National Pollutant Discharge Elimination System (NPDES) permit. These permits are issued under authority of the Clean Water Act[38] and implementing regulations[39] by the Environmental Protection Agency (EPA) or a state agency delegated permit program authority by the EPA.

According to the regulatory constraints report,[40] fish farmers argue that EPA regulations are unduly restrictive in that they fail to distinguish between biodegradable wastes produced by fish hatcheries, and chemical wastes produced by industry. They also complain that the beneficial effects of fish waste nutrients is not considered, nor the flushing effect of tidal waters in some locations.

In making a determination that an aquaculture facility is a "point source" requiring a NPDES permit, regulations require an on site inspection of the location and quality of receiving waters, holding, feeding and production capacity of the facility, and, the quantity and nature of the pollutants reaching waters of the U.S.[41] Aquaculture applicants must provide this information to the EPA and must report quantitative data on effluent characteristics of the same kind as manufacturing, commercial, mining and silvicultural discharges.

Section 404 of the Clean Water Act authorizes the U.S. Army Corps of Engineers to regulate the discharge of dredged or fill materials into waters of the U.S.[42] Curiously, the guidelines for issuing these permits are Section 404 (b)(1) guidelines issued by EPA. Responsibility for the program is thus split between the Corps and EPA. Any construction of dams or dikes in navigable waters; any activities that alter the course, condition, location or capacity of navigable waters; and, all transportation of dredged materials for dumping into ocean waters require permits. A privately owned waterway might come within the legal definition of "navigable waters of the U.S."[43]

In addition to the Corps and EPA, the Fish and Wildlife Service of the Department of the Interior and the National Marine Fisheries Service of the Department of Commerce may be given the opportunity to review and comment on dredge and fill applications where fish and wildlife resources could be affected.

Facility/Hatchery Management and Processing

Having obtained the necessary land use, siting, water use and pollution permits, and having complied with applicable federal, state and local business and tax regulations, the fish farmer must consider the spectrum of public health and sanitation regulations which affect the heart of an aquaculture operation. These regulations might present major obstacles arising from biological or chemical contamination of water, pharmaceutical residues from commercial feeds or water additives, and diseases in the fish.[44] Pesticides, herbicides and chemicals used for predator control in aquaculture operations are strictly regulated at each level of government. Even so, as a recent report of the NJDEP Office of Cancer and Toxic Substances[45] makes painfully evident, fish represent various risks to consumers since they directly reflect the character of the environment from which they originate.

An absolute need for quality aquaculture raised fish products raises issues of fish disease control for the fish farmer. The Food, Drug and Cosmetic Act[45] (FDCA) requires that drugs or chemicals used on fish must be registered and approved by the Food and Drug Administration (FDA) of the Department of Health and Human Services. Approval of such chemicals and pharmaceuticals requires a rigid and highly specific FDA certification process; one which might cost years of research and millions of dollars. Within the FDA, the Bureau of Veterinary Medicine, the Bureau of Foods and the Bureau of Toxicology review applications, and for aquaculture purposes the Fish and Wildlife Service would also become involved in the review process.

Because of the probable high cost in time and money involved in obtaining FDA certification, and because aquaculture operations present a relatively small market, private industry to date has not made the investment in research and development which would provide an adequate battery of pharmaceuticals and chemicals for the fish farmer.

Another facet of the FDA drug registration process which presents a constraint to the aquaculture industry is the fact that the process applies to the use of a drug, not to the drug itself. The current system requires separate studies and reregistration for use on separate species. A drug approved for use on trout would not automatically be approved for use on salmon. The constraining effect of this process is obvious, but the FDA has consistently rejected proposals that some drugs be given "blanket approval" for a variety of species and uses.

The net result presents a dilemna to the fish farmer: by not using unregistered chemicals, valuable fish stock will be lost to parasites and disease; by using unregistered chemicals with known therapeutic value, the aquaculturist risks being in violation of federal regulations.

Food additives must likewise be cleared for safety by the FDA before use in processing, packaging, transporting or holding fish products. Adulterating or misbranding any food or drug is illegal,[47] and for purposes of the FDCA a food is adulterated if it " . . . contains any poisonous or deleterious substance which may render it injurious to health."[48] Proving the safety of a given substance also requires extensive testing.

Finally, since an aquaculture facility is normally an integrated operation, the prospective aquaculturist should be aware that both state health agencies and the FDA specify design and construction requirements of processing plants as well as operational procedures to insure a safe and wholesome product.

Fish and Fisheries Management

The government agencies chiefly responsible for development and implementation of fisheries management programs are the National Marine Fisheries Service (NMFS) of the U.S. Department of Commerce, the Fish and Wildlife Service (FWS) of the U.S. Department of Interior, and the fifty corresponding state agencies. These agencies are responsible for the conservation and maintenance of healthy stocks of fish in health habitats for commercial and recreational fishing.

Congressman John F. Lacey of Iowa introduced the original Lacey Act of 1900[49] to assist individual states in protecting wildlife, chiefly bird and animal species, from illegal interstate traffic. The act provided for federal jurisdiction over such species, moved beyond originating state jurisdiction. The Black Bass Act of 1926,[50] based on the same philosophy, was ultimately expanded to cover all species of fish. The illegal movement of fish across state (and national) boundaries was identified by Congress as an increasing problem involving tremendous illegal profits,[51] and the Lacey Act amendments of 1981 consolidated and strengthened the applicable laws.

The Lacey Act is aimed at the protection of wildlife, the restriction of importation of non-indigenous (potentially harmful) species, and the control of animal, bird and fish diseases and parasites. It is not aimed at constraining the aquaculture industry, although aquaculturists are subject to the law. The importation, exportation and transportation of wildlife is restricted, both by the federal Act and by applicable laws and regulations of the individual states. The term "fish or wildlife" means any " . . . fish, . . . whether or not bred, hatched, or born in captivity" which is normally found in a wild state, " . . . and including any part, product, egg or offspring thereof."[53] Anyone who knowingly receives, acquires or purchases any prohibited species is liable under the act for a civil penalty of up to $10,000[54] or criminal penalties of up to $20,000 or five years in prison or both.[55]

The potential aquaculturist must therefore make certain that the necessary state permits; hatchery permits, stocking permits, etc., are obtained before acquiring eggs, fingerlings or brood stock which travels in interstate commerce.

CONCLUSIONS

It is not necessarily individual laws or regulations that constrain the aquaculture industry, it is the enormous weight of numbers of regulations. Movement toward a coordinated federal program is taking place, particularly since passage of the National Aquaculture Act of 1980. At the very least this Act has resulted in the identification and compilation of specific regulatory constraints. The greatest number of regulations exist at the state level however, and few states have even considered a coordinated approach to aquaculture regulation. There is also some movement in that direction on the part of individual states, but it is too early to accurately measure the success of such programs.

For the indefinite future the potential aquaculture entrepreneur is likely to be engaged in a time consuming and expensive regulatory process.

REFERENCES

1. U.S. Department of Commerce, U.S. Ocean Policy In The 1970's: Status and Issues, Washington, D.C., Government Printing Office, October 1978, III-18.

2. 16 U.S.C. 1801-1882, P.L. 94-265, as amended.

3. U.S. Department of Commerce, *supra,* n. 1 at II-2.

4. *See*, Saltonstall-Kennedy Act, 15 U.S.C. 713c-3.

5. Watson, A. Shaw, Aquaculture and Algae Culture: Process and Products, Noyes Data Corporation, Park Ridge, N.J., 1979, P.1.

6. U.S. Comptroller General, The U.S. Fishing Industry-Present Condition And Future Of Marine Fisheries, Washington, D.C., Government Printing Office, 1976, (107 pp.) p.ix.

7. National Fishery Law Symposium, Proceedings of First Annual Meeting: Advising The Fishing Industry, October 14-16, 1982, University of Washington School Of Law Foundation, p. A-26.

8. U.S. House of Representatives, Committee on Merchant Marine and Fisheries, Kent Price, Aquaculture Hearings before Subcommittees on Fisheries and Wildlife Conservation and the Environment and Oceanography, Washington, D.C., Government Printing Office, March 15, 1977.

9. Watson, A. Shaw, *supra,* n.5 at 2.

10. 16 U.S.C.A. §2801 *et seq.*

11. 1980 U.S. Code Cong. and Adm. News, p.2878.

12. 16 U.S.C.A. §2803.

13. 16 U.S.C.A. §2803(b)(5)

14. 16 U.S.C.A. §2803(e)(6)

15. 16 U.S.C.A. §2808(a)

16. U.S. Fish and Wildlife Service, "Aquaculture In The United States: Regulatory Constraints", Aspen Research And Information Center, Report No. 14-16-009-79-095.

17. U.S. Fish and Wildlife Service, *supra,* n.16 at III-3.

18. U.S. Fish and Wildlife Service, *supra,* n.16 at IV-1.

19. Telephone conversation with Bille Hougart, Aquaculture Coordinator, U.S. Department of Agriculture, January 5, 1983.

20. Haar, Charles, M., *Land-Use Planning: A Casebook On The Use, Misuse, And Re-Use Of Urban Land,* Little, Brown and Company, Boston-Toronto, 1976, xi.

21. *Village Of Euclid v. Ambler Realty Co.,* 272 U.S. 365, 47S.ct. 114, 71L.Ed. 303 (1926).

22. The area of takings remains one of the most confused and confusing areas of land-use law. For an examination of taking issues *see, e.g.,* Grad, Frank P., Rathjens, George W., Rosenthall, Albert J., *Environmental Control: Priorities, Policies, And The Law,* Columbia University Press, New York and London, 1971.

23. Coastal Zone Management Act, 16 U.S.C. 1451 *et seq.*

24. P.L. 92-583, Sec. 304(1).

25. As of September 30, 1982 the Office of Coastal Zone Management had approved 28 programs. U.S. Department of Commerce, *Coastal Zone Management,*

NOAA, 3300 Whitehaven Rd., Washington, D.C. 20235.

26. Waterfront Development Law, N.J.S.A. 12:5-3.

27. Wetlands Act, N.J.S.A. 13:9A-1 *et seq.*

28. Coastal Area Facility Review Act, N.J.S.A. 13:19-1 *et seq.*

29. 16 copies including one reproducible transparency.

30. N.J.S.A. 13:19-3 (c)(2).

31. N.J.A.C. 7:7D-2.4.

32. N.J.S.A. 12:3.1. *et seq.* There are three kinds of tidelands instruments: a grant conveys full ownership; a lease conveys use of property for a fixed number of years (usually issued for projects involving solid fill), and a license also allows use of property for a fixed number of years (normally 10 or less for residential docks and piers).

33. N.J.S.A. 12:3-23.

34. *See e.g.*, *The Port and Waterways Safety Act*, 33 U.S.C. §1221 *et seq.*; 46 U.S.C. §391a.; The Rivers and Harbors Act of 1899, 33 U.S.C. §§401, 403, 407.

35. Submerged Lands Act, 43 U.S.C. 1301-1343, Subchapter II.

36. Outer Continental Shelf Lands Act, 43 U.S.C. 1301-1343, Subchapter III.

37. *E.g.* N.J.S.A. 12:3-23.

38. The Clean Water Act, 33 U.S.C. §1321.

39. 40 C.F.R. 122 (Appendix C.)

40. U.S. Fish and Wildlife Service, n.16 at II-6.

41. 40 C.F.R. 122.55(c).

42. 33 U.S.C. §1344.

43. *U.S. v Kaiser Aetna,* 408 F. Supp. 42 (D. Hawaii 1976).

44. Watson, A. Shaw, n.5 at 297.

45. NJDEP, *PCB's In Fish: A Comprehensive Survey,* Office of Cancer and Toxic Substance Research, November, 1982.

46. Food, Drug and Cosmetic Act, 21 U.S.C. 346a.

47. 21 U.S.C §3316.

48. 21 U.S.C. §342a.

49. Lacey Act, 16 U.S.C. 667e.

50. Black Bass Act, 16 U.S.C. 851.

51. U.S. Congress, House, *Lacey Act Amendments of 1981,* House Report 97-276 to accompany H.R. 1638, 97th Congress, 1st sess., 1981.

52. P.L. 97-79, 16 U.S.C. §3371 *et. seq.*

53. 16 U.S.C. §3371(a).

54. 16 U.S.C. §3373(a).

55. 16 U.S.C. §3373(d).

Marketing Striped Bass

Dave Swartz

This paper will focus on the food fish market opportunities open to the striped bass culturist. Historic information is provided on general prices and supply of striped bass along with detailed information on market channels, desired product forms and marketing services. In future work the live markets for striped bass and hybrids will be explored in greater detail.

Many aquaculture projects today have proven to be technologically sound, yet they may not be currently economically viable. The reasons for a lack of profitability are associated with high production costs, lack of available financing, and improper attention to marketing or legal restraints. To be economically feasible a venture must be able to produce a product at a reasonable cost and sell it at a competitive price. To get the best price a producer has to be familiar with the marketing of his product. The purpose of this paper is to provide producers of striped bass with insights into the marketing of their product in the Mid-Atlantic region.

GENERAL MARKET OUTLETS FOR STRIPED BASS

The marketing of cultured striped bass and hybrids offers much opportunity for varied product forms and market channels. The producer of striped bass has greater flexibility than traditional suppliers of food fish products, and he will want to exploit the comparative advantages he enjoys over traditional suppliers. These advantages relate to the control over a number of factors, including size, quality, product form and time of harvest. In order to utilize market conditions to best advantage, a producer will want to sell his products during the off-season for commercial fishing, when prices are generally higher due to a lack of supply. Similarly, it will be of advantage to grow the fish to a size below the minimum allowed in a capture fishery to satisfy specific market demands. The producer also has more control over quality and product form. Figure 1 outlines the general product forms and outlets for striped bass.

In general, the market can be partitioned between live and food fish sales. Live fish can be sold to fish-out operations where consumers pay to harvest fish from a pond stocked with fish, either by the pound ($1.50-2/lb) or through a general admission fee. Also generally supplied are rental equipment, bait, food services, ice and restroom facilities. This market outlet is popular for trout and salmon. Similarly, striped bass fishouts will likely meet with broad consumer acceptance and interest, especially as the availability of wild stocks become scarce. Live striped bass may also be sold to power plants for mitigation purposes or to state programs for fish stocking. More recently the marketing of live adult striped bass has proven successful among the oriental and kosher markets of California and New York, respectively. The premium prices offered by these markets (January 1983, $6/live lb.)[1] provide the incentive to expand this live market to restaurants and other markets desiring the freshest of products. Also available to the producer of cultured fish are traditional food fish markets, such as wholesale, restaurant, retail and consumer markets.

HISTORICAL INFORMATION ON MARKETS FOR STRIPED BASS

While current commercial harvests of striped bass are less than 30% of the peak 1973 harvests of approximately 15 million lbs., they are above the levels of the early 1930's. Since 1930, harvests have grown to a peak in the early 70's and fallen

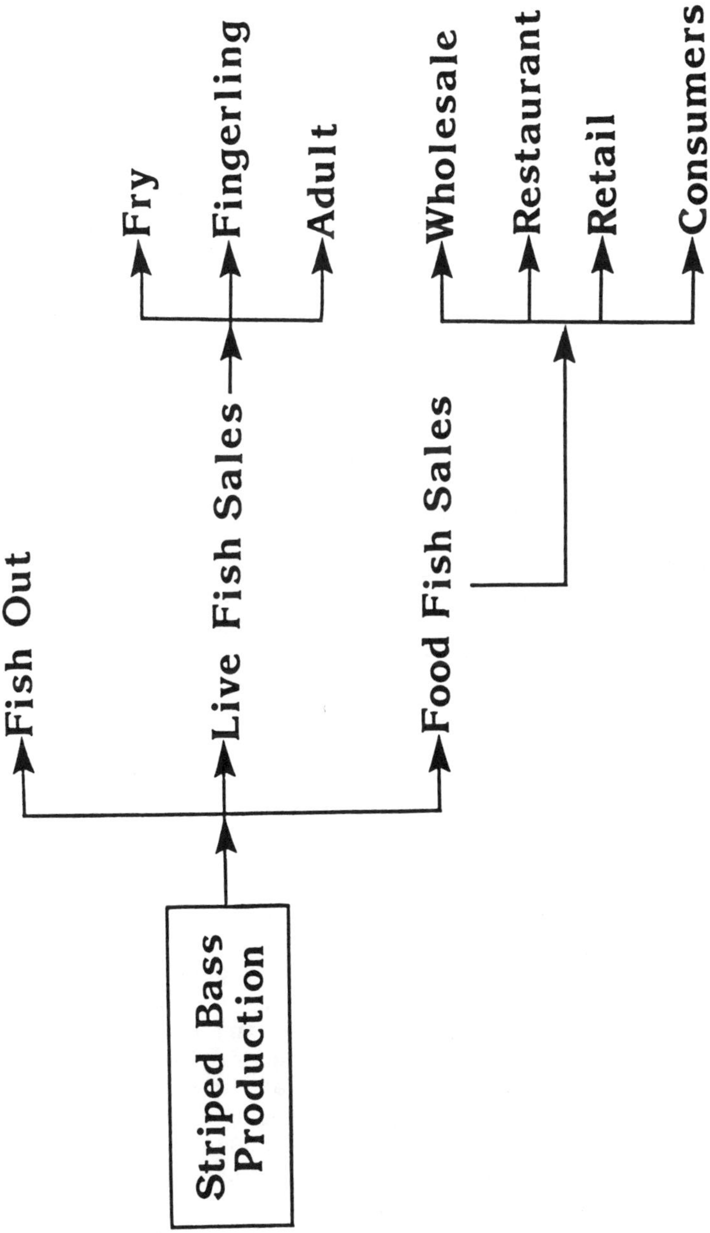

Figure 1. General market outlets for striped bass.

until the late 70's and remained stable through the late 70's and early 80's. See Figure 2 for a plot of striped bass landings since 1930.

The real price[2] of striped bass peaked several times over the past 50 years: in the early 30's, early 40's and mid 70's. Associated with these increases in prices are low levels of striped bass harvests. In general, when the harvests of striped bass are low we would expect the price to be high (relative to when harvests are high).

Other factors would also be expected to affect the price of striped bass. The level of disposable personal income, for example, could also influence prices. When real disposable incomes are high we would expect prices to be higher than when incomes are low. Also affecting striped bass prices are the availability and prices of substitute products such as those of other finfish or other sources of protein. Finally, general changes in the tastes and preferences of individuals can affect long-run price levels. In the past decade the per capita consumption of fish products has increased by 25 percent. This increase in personal consumption over the past decade has been associated with a growing concern for health through diet and exercise. Low calories, cholesterol and fat levels in fish make it a desirable food for health-conscious individuals. Adriance (1982) has explored in detail the relationship of striped bass prices to those factors discussed above.

If landings of striped bass are examined on a monthly basis, we can observe that some interesting patterns present themselves. In Figure 3 monthly Maryland landings of striped bass are graphed for the years 1973-76. Noteworthy is the seasonality of these landings, which peak in spring and are low in summer, fall and winter. Unlike Maryland, New England's striped bass harvests peak in the summer months. For Mid-Atlantic states north of Maryland, harvests tend to pick up in summer and early fall. The seasonal nature of supply makes it necessary for consumers, restaurants and others to obtain striped bass from different areas when local re-

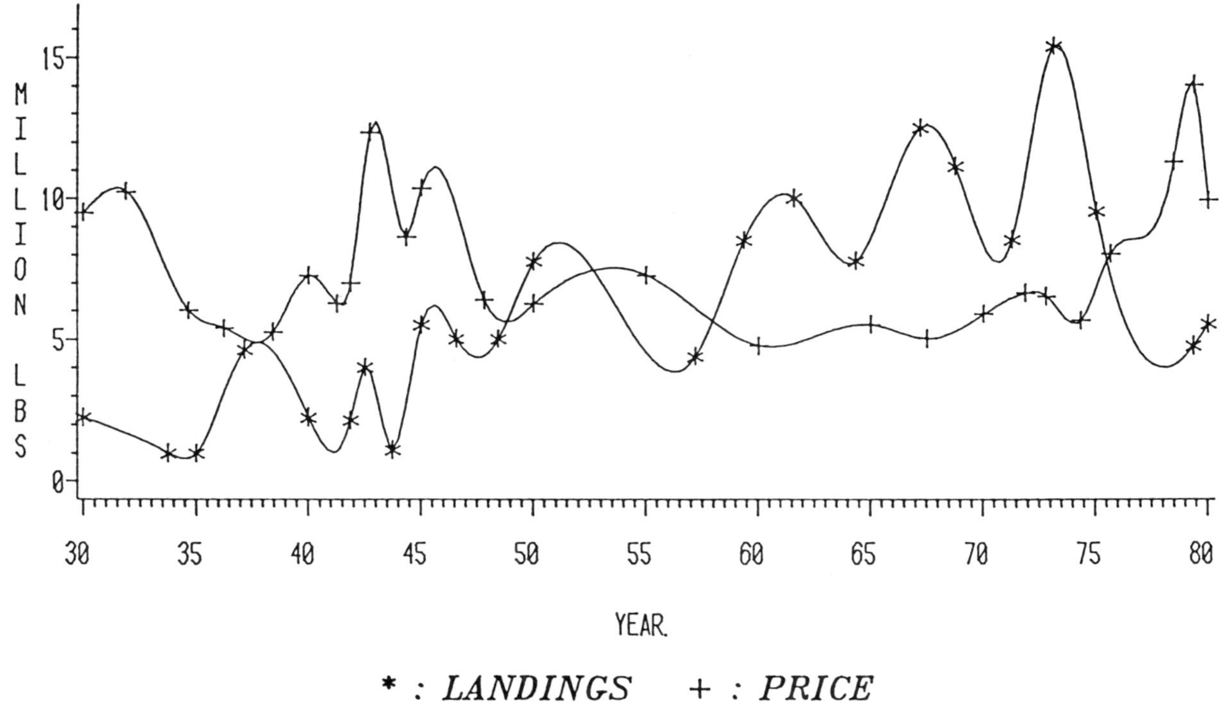

Figure 2. Atlantic Coast landings and adjusted prices for striped bass.

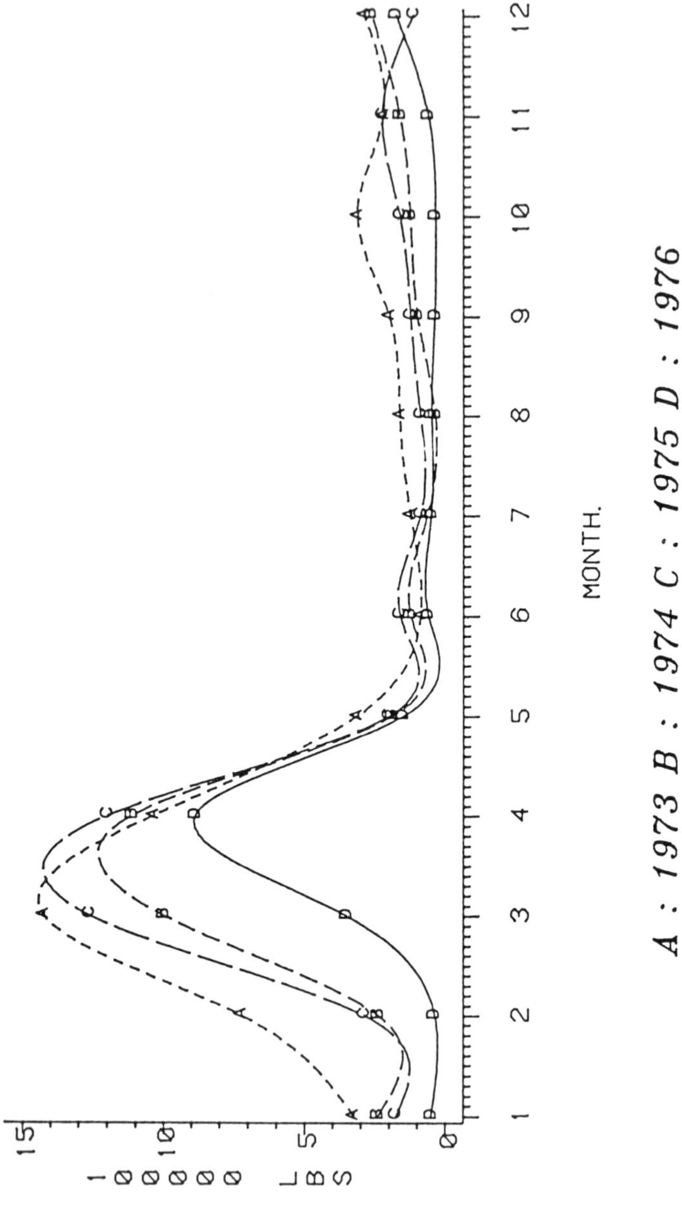

Figure 3. Maryland striped bass landings, 1982.

gional supplies are scarce. This scarcity, along with additional transportation costs, tends to drive local prices up when local supplies are scarce. Taking this factor into account, an aquafarmer performing direct marketing should focus his efforts on those regions demonstating scarcity effect.

Figure 4 provides recent prices observed at Fulton Wholesale Market in New York City. These prices are available through the daily and weekly summaries of seafood prices published in the New York Market News (Green Sheets) available through the National Marine Fisheries Service. Prices in New York City tend to influence prices throughout the Mid-Atlantic. Wholesalers and others separated from centralized markets turn to centralized markets like Fulton and Baltimore's Lexington Market for guidance in setting their prices. Wholesale prices during 1982 have tended to fall between $2-3/lb., with ex-vessel prices between $1-2/lb. and retail prices between $3-4/lb. (and up to $6/lb).

MARKET CHANNELS AND PRODUCT FLOW FOR STRIPED BASS AS FOOD FISH

Generalized Marketing Channels

Prior to a specific discussion of market channels for striped bass it would be beneficial to review briefly a generalized diagram of marketing channels for seafood. Figure 5 is such a generalized diagram which outlines the product flow from producer to the consumer. Between these market levels are many of the middle men who add value to the product or provide a necessary marketing service. The producer generally sells to a buyer at the dock who may work independently or directly for a processor or a larger wholesaler. Eventually the product is distributed to markets such as restaurants, grocery stores or institutions. In the following specific diagrams of striped bass marketing channels the different market levels presented in the simplified diagram are represented.

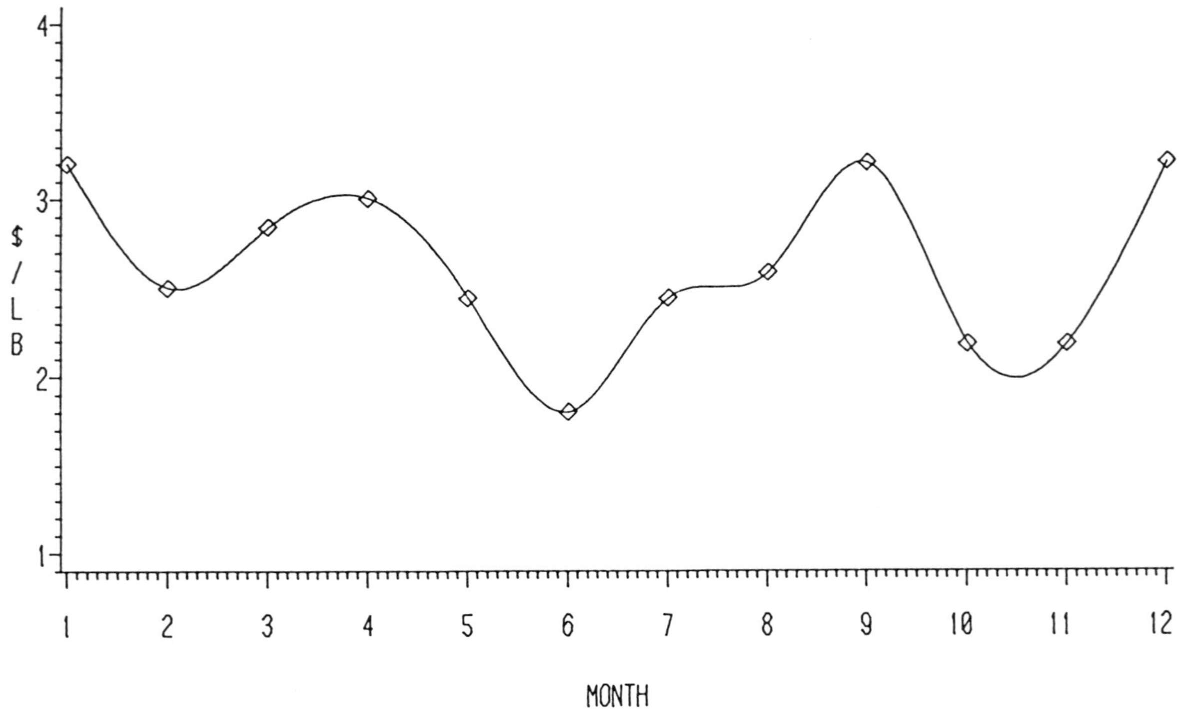

Figure 4. Fulton Market prices for striped bass, 1982.

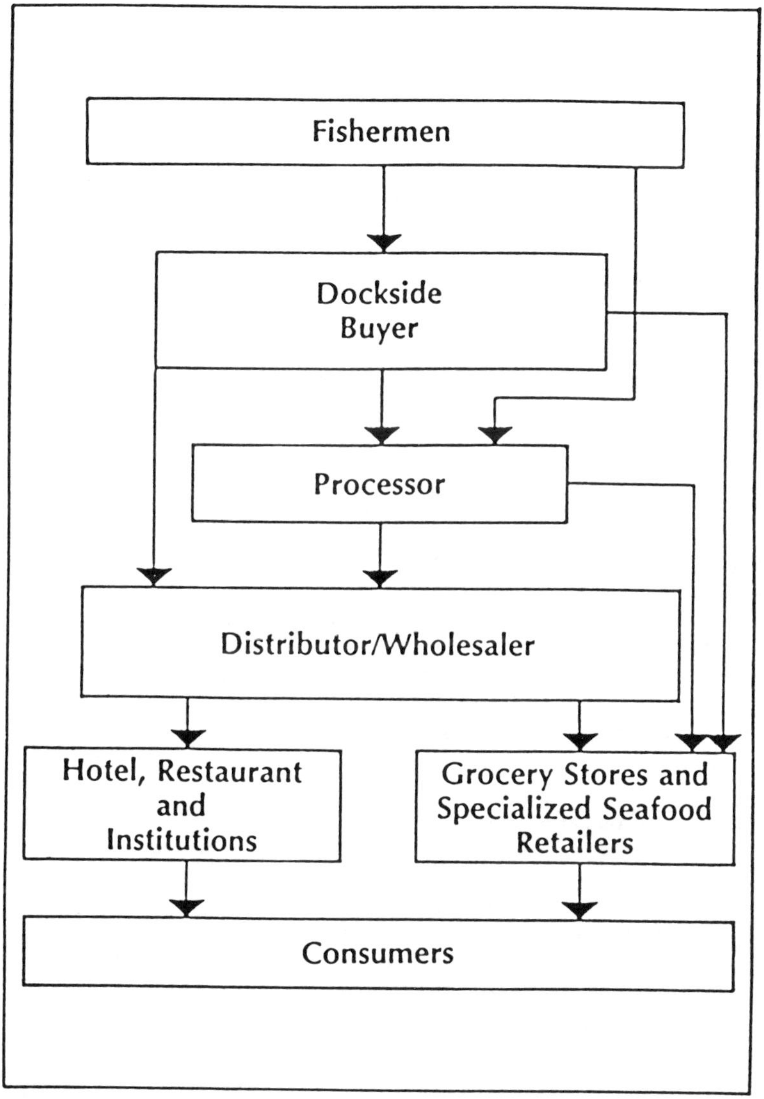

Figure 5. Generalized diagram of major marketing channels for domestic seafood products.

Market Channels for Chesapeake Area

As in the simplified market channel diagram, Figure 6 provides an overview of the product flow through the Maryland market. Starting with harvesters, striped bass are purchased at the dock by wholesalers. The food fish eventually reaches the various higher level markets for food fish including restaurants and retail outlets. Most of the striped bass harvested in Maryland are taken by northern Bay gill netters. Almost all of those fish are sent to the Lexington (Baltimore) or the Fulton (NY City) wholesale markets. In general, larger fish are sent to the Fulton Market while the smaller fish remain in Baltimore. The following market summary sections go into greater detail on this point.

East Coast Market Channels

The market for striped bass entails a complex array of producers, wholesalers and retailers. Figure 7 provides a product flow diagram of the entire east coast market for striped bass. The previous diagram for Maryland markets is embedded in this comprehensive summary of striped bass marketing flow. Interesting to note is that in addition to vertical pathways between market levels in a region or among regions there are also horizontal pathways between regions at a given market level. As product becomes scarce, the role of distant wholesalers becomes more important in satisfying regional demands during periods of scarcity. Of course, as the product passes through each middleman the market margin tends to grow quite rapidly. This has resulted in a situation where retail prices above $6 are becoming common. The aquafarmer has the option of entering these markets at whatever level he chooses, provided he can supply the quantity of fish and the marketing services to satisfy the requirements of the specific market channel. If a producer is positioned to meet these requirements through the integration of different middlemen

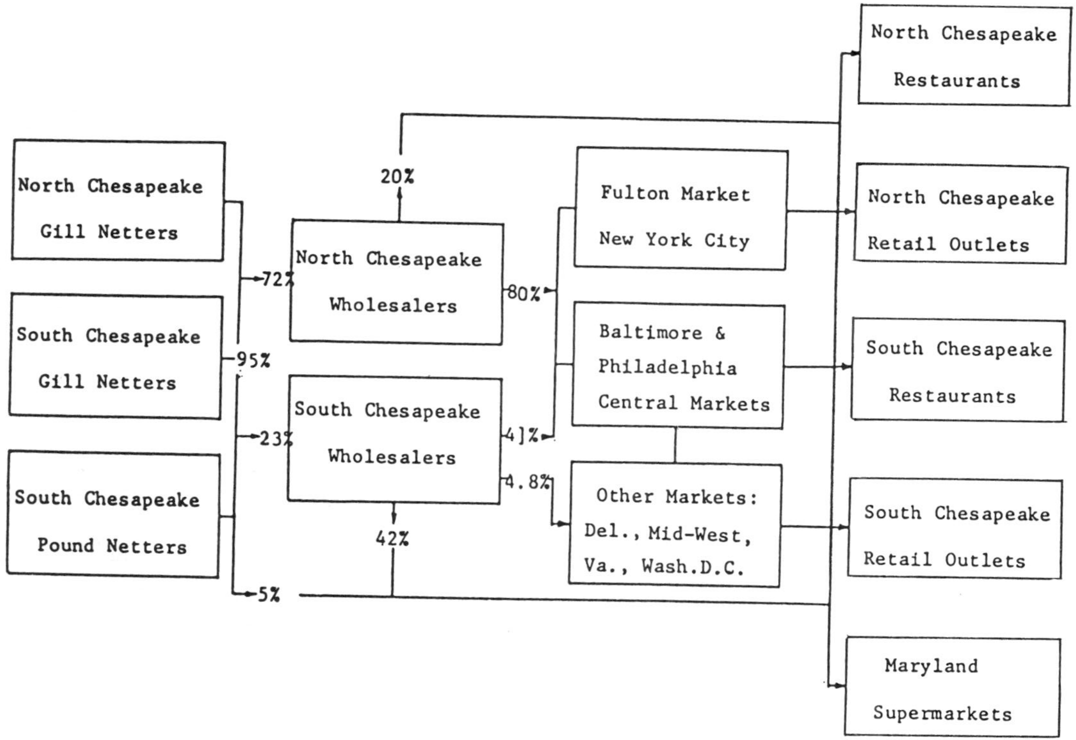

Figure 6. Maryland striped bass flow, 1980, by percent.

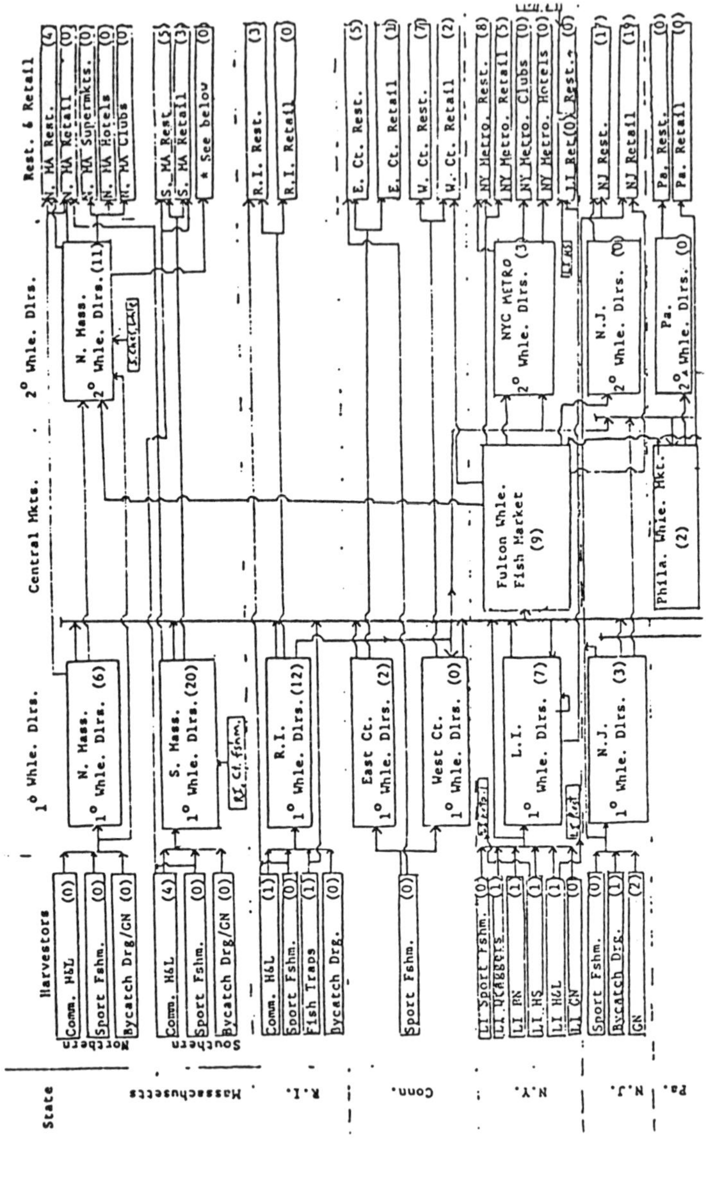
State
Harvestors
1° Whle. Dlrs.
Central Mkts.
2° Whle. Dlrs.
Rest. & Retail
Massachusetts
Northern
Southern
Comm. H&L (0)
Sport Fshm. (0)
Bycatch Drg/GN (0)
N. Mass. 1° Whle. Dlrs. (6)
N. Mass. 2° Whle. Dlrs. (11)
N. MA Rest. (4)
N. MA Retail (0)
N. MA Supermkts. (0)
N. MA Hotels (0)
N. MA Clubs (0)
Comm. H&L (4)
Sport Fshm. (0)
Bycatch Drg/GN (0)
S. Mass. 1° Whle. Dlrs. (20)
S. MA Rest (5)
S. MA Retail (3)
* See below (0)
R.I.
Comm. H&L (1)
Sport Fshm. (0)
Fish Traps (1)
Bycatch Drg. (0)
R.I. 1° Whle. Dlrs. (12)
R.I. Rest. (3)
R.I. Retail (0)
Conn.
Sport Fshm. (0)
East Ct. 1° Whle. Dlrs. (2)
West Ct. 1° Whle. Dlrs. (0)
E. Ct. Rest. (5)
E. Ct. Retail (1)
W. Ct. Rest. (7)
W. Ct. Retail (2)
N.Y.
LI Sport Fshm. (0)
LI Draggers (1)
LI PN (1)
LI HS (1)
LI H&L (1)
LI GN (0)
L.I. 1° Whle. Dlrs. (7)
Fulton Whle. Fish Market (9)
NYC METRO 2° Whle. Dlrs. (3)
NY Metro. Rest. (8)
NY Metro. Retail (5)
NY Metro. Clubs (0)
NY Metro. Hotels (0)
LI Ret(0) Rest. (0)
N.J.
Sport Fshm. (0)
Bycatch Drg. (1)
GN (2)
N.J. 1° Whle. Dlrs. (3)
N.J. 2° Whle. Dlrs. (0)
NJ Rest. (17)
NJ Retail (19)
Pa.
Phila. Whle. Mkt. (2)
Pa. 2° Whle. Dlrs. (0)
Pa. Rest. (0)
Pa. Retail (0)

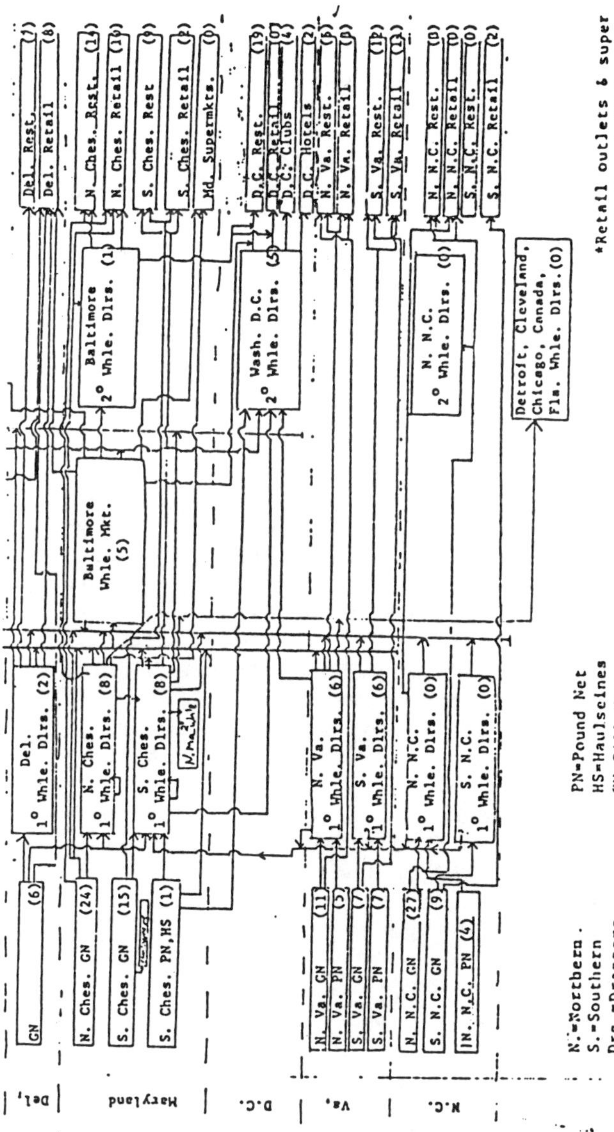

Figure 7. Striped bass marketing flow, 1980.

functions, then substantial margins can be realized. However, prior to embarking on such a path it is advisable to compare the costs and benefits of integration. One cost is the lost opportunity, for a manager, in production while time is spent on marketing. For those who are not marketeers, it may be advisable to leave that job to the pros--but remember that you may pay dearly for the service.

MARKET SUMMARY FOR RETAIL, RESTAURANT AND WHOLESALE LEVELS

Retail Market Summary

This section provides an overview of the retail market for striped bass from New York to North Carolina. The information provided is based on a lengthy survey conducted by the University of Maryland. Table 1 includes a listing of the source of the product--whether it originates from wholesalers or fishermen. It lists the size preference of the market and the retail price (in the round). It also provides the range in volume of products handled by the retailer during the survey year. It is clear that the fish size is an important factor which varies among regions. In the northern Atlantic region (New York, New Jersey and Delaware) there exists a preference for larger fish for the preparation of fillets. The price for larger fish is also greater per pound in these regions. In the southern regions (Maryland, Virginia, North Carolina) there exists a preference for the smaller pan-sized fish.

Restaurant Market Summary

One premium market for striped bass exists with the restaurants (see Table 2). Direct marketing to restaurants can net large margins for producers willing to take time to open this marketing channel. Direct marketing to restaurants is common in resort areas where the fishing season corresponds to the vacation season. However, when the distance

Table 1. Retail Market Summary

	General Brought From	Size Preference	1980 Retail Price/lb.	Total 1980 Striped Bass Handled (lbs.)
New York	Fulton Mkt.	2-30 lbs.	$3.00-6.50	50 -15,000
New Jersey	Fisherman & Wholesalers	1-35 lbs.	$1.45-6.00	0 -2,500
Delaware	Wholesalers	1-15 lbs.	$1.25-3.69	400 -4,800
Maryland	Wholesalers	3/4-4 lbs.	$ .85-4.00	1,800-20,000
Northern Virginia	Wholesalers	1-4 lbs.	$1.00-3.00	150 -50,000
Southern Virginia	Wholesalers	1-3 lbs.	$ .95-3.00	100 -10,000
North Carolina	Fisherman & Wholesalers	1 1/2-4 lbs.	$1.59-1.69	2,500-9,000

Table 2. Restaurant Summary

	General Brought From	Size Preference	Average Entree Price	Total 1980 Striped Bass Handled (lbs.)
New York	Purveyors	2 1/2-20 lbs.	$6.95-13.00	30 -4,800
New Jersey	Wholesalers & Fishermen	1-40 lbs.	$5.50-11.00	30 -2,000
Delaware	Wholesalers	2-45 lbs.	$6.95-12.95	100-10,000
Northern Maryland	Wholesalers & Purveyors	1-14 lbs.	$4.50-14.00	200-10,000
Southern Maryland	Fishermen & Wholesalers	1-3 lbs.	$5.75- 9.95	24 -20,000
Washington, D.C.	Wholesalers	1-10 lbs.	$6.00-16.00	50 -15,000
Northern Virginia	Wholesalers	1-20 lbs.	$5.25-13.95	50 -7,500
Southern Virginia	Wholesalers	1-20 lbs.	$4.75-12.50	100-1,500
North Carolina	Wholesalers & Fishermen	2-4 lbs.	$5.25- 9.00	-

from the market grows and when harvesters can no longer supply product due to closed seasons, then restaurants tend to depend on wholesalers and purveyors (a specialized supplier of restuarants) to meet their product needs.

An aquafarmer, by meeting the annual supply and high quality standards required by restaurants, will likely meet with success in pursuing this market channel. Some restaurants, according to the survey, handle large amounts of fresh product. The regional size preferences exist to a lessor degree than retail markets, as restaurants (especially French) prefer to serve fillets from large striped bass.

Wholesaler Market Summary

The survey of wholesalers has revealed that wholesalers tend to specialize in a narrow range of products or generalize by supplying a broad range of seafood products (see Table 3 for a summary). In some cases these wholesalers are large and move a large percentage of the available striped bass. The largest of these wholesalers are located in New York and the Chesapeake area.

MARKETING SERVICES REQUIRED AND PRICE DETERMINATION IN STRIPED BASS MARKETS

Marketing Services Required

In order to enter certain marketing channels for striped bass an aquafarmer should be prepared to provide or contract for certain marketing services. Figure 8 provides an overview of the importance of different marketing services to various market channels, such as, consumers, hotels and restaurants and food chains. Services summarized include heading and gutting, scaling, filleting, freezing and chilling. The size of the darkened circle in Figure 11 suggests the importance of the service to the corresponding market channel.

To market directly to consumers, a producer should be able to scale, clean and head the fish.

Table 3. Wholesaler Summary

	Generally Brought From	Ships Bass To:	Pounds of Striped Bass Handled in 1980
New York Long Island	Fishermen	Local Restaurants Local Wholesalers Retailers	5,000- 50,000
New York New York City	Wholesalers Fishermen	Local & Non-local Retailers, Wholesalers, Restaurant Suppliers	100,000-156,000
New Jersey	Fishemen	Fulton Market	1,300- 10,000
Delaware	Fishermen	Non-local Wholesalers	1,500- 1,600
Northern Maryland	Wholesalers Fishermen	Restaurants, Wholesalers, Processors	5,000-700,000
Southern Maryland	Fishermen	Wholesalers, Restaurants Fulton Market	2,000-500,000
Washington, D.C.	Non-local Wholesalers	Retailers, Wholesalers Restaurants	3,000-100,000
Northern Virginia	Fishermen Wholesalers	Wholesalers, Restaurants Retailers, Fulton Market	100-375,000
Southern Virginia	Fishermen Wholesalers	Restaurants, Retailers Fulton Market	100- 15,000

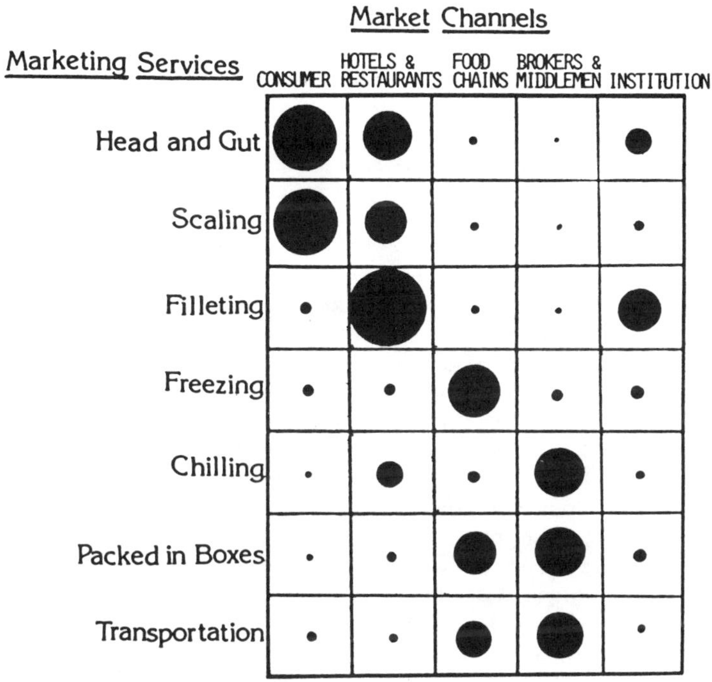

Figure 8. Importance of marketing services to market channels (from Yamashita 1981).

Hotels and restaurants require similar services to consumers, with the addition of filleting and chilling of the product. Food chains for the most part require a frozen product, and to a smaller degree fish iced and packed in the round. Transportation of the product to food chains is also required. Middlemen prefer fish iced and packed in the round. Transportation is also important, but some middlemen arrange for transportation. Institutional markets require processed fish--cleaned and headed or filleted.

Price Determination and Differentiation

The forces of supply and demand play an important role at centralized markets in the process of price determination. Remote markets depend to a great extent on centralized market prices for guidance in setting their own prices. These centralized market prices are published by the National Marine Fisheries Services in the Market News Blue and Green Sheets. Figure 9 provides an overview of the importance of centralized market prices to price determination in remote markets. Again, the size of the darkened circle corresponds to the importance of a factor such as negotiation or centralized market price to price determination. Negotiation over price plays an important role in Maryland, while in Virginia sellers depend more on centralized market prices for guidance in setting prices.

Prices of fish products in Maryland and Virginia are differentiated according to the size of the fish, the area it was landed, the lot size or volume of fish sold and the gear type used to harvest the fish. The area landed and gear type relate to the quality of fish. Some gear tends to damage the harvested fish more than other types. An aquafarmer will have a great deal of control over these factors so that premium prices can be obtained for these products.

SUMMARY AND CONCLUSION

To insure the development of a viable aquaculture industry, it will be necessary for industry, government and universities to form a partnership to help address problems of concern and to find satisfactory solutions. There are a number of impediments to overcome before aquaculture can secure a place as an important contributor to the nation's food supply. This paper has not, for example, addressed legal or institutional impediments to marketing. There is great opportunity for an aquafarmer to enter different market channels for striped bass and to secure premimum prices for his product. Before maximum advantage can be taken of

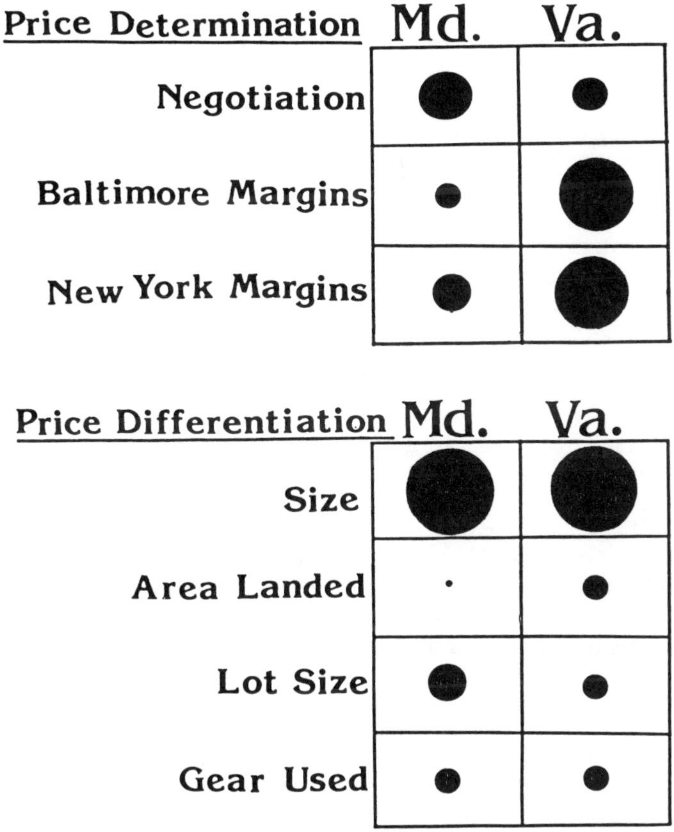

Figure 9. Methods of price determination and differentiation in Maryland and Virginia.

this opportunity, states will have to ammend laws governing restrictions on the import and marketing of fish during off-seasons and of fish under or over regulated sizes. In Maryland, for example, it is illegal to market striped bass below 14 inches long in size. Laws such as these need to be examined and ammended to reflect the true intent of the law relative to cultured--as opposed to wild--striped bass. Once this is accomplished, aquaculture will advance one step further toward becoming an important supplier of much-needed food products.

END NOTES

[1]Note that a live pound includes water weight which is lost if the fish is butchered for food fish markets.

[2]Real price refers to the price of striped bass adjusted or deflated for inflation by the formula: P/WPI X 100 = Adjusted Price, where WPI = Wholesale Price Index and P = Price.

REFERENCES

Adriance, J.G. 1982. Market Model for the North Atlantic Commercial Striped Bass Industry. Master's Thesis, University of Maryland.

Yamashita, V.F. 1981. Marketing of Striped Bass in the Chesapeake Bay States of Maryland and Virginia. Master's Thesis, University of Maryland.

An Optimist's View of the Future

Joseph P. McCraren

Following some 20 years with the Fish and Wildlife Service, I fully admit to possessing a crystal ball with enough cracks in its surface to render it less than useful to an aspiring newcomer to the fisheries profession. One of the deepest and ultimately most significant irregularities in its surface occurred when I was stationed as a biologist at an old and venerable hatchery located some 10 miles outside of a town with a population of 1500. The town was Tishomingo, Oklahoma, situated in the southcentral portion of the state. To say that Tishomingo was isolated would be an understatement. I think a young, summer temporary employee summed it up well when--upon arriving for duty and having visually inspected his new home for the next 3 months--he turned to me and drolly stated, "Rather rural, isn't it?"

Rural? Yes, but Tishomingo has a productive pondfish program that exists to this day. I was reminded of Tishomingo and the young man because it was there that I was first introduced to the striped bass, some 12 or 13 years ago. Reference to the inordinately cracked crystal ball stems from the fact that I was less than inspired with our entrance into culturing striped bass. There were other species to continue working with, such as black bass, catfish, and common sunfish the "bread and butter" species of the pondfish hatchery for which a multitude of cultural problems existed and that remain unresolved today.

But the "striper" was it. It was the "comer." I resisted. I argued with my staff and supervisors. I insisted upon continuing to place emphasis upon our native, traditional species. The striper was analogous in my mind to a wildlife management biologist's umpteenth introduction of another game bird from perhaps the far reaches of the Gobi Desert. Right? Wrong!

Word of success in the Southeast kept coming as steadily as the fry shipments from Monck's Corner. And we, more importantly I, began to catch the fever. Never mind the fact that fry invariably arrived in Dallas at midnight, had to be transported to the hatchery, tempered and if old enough "stocked out" in the ponds--usually sometime just before breakfast. And never mind that our first impression of the fry was that they were not unlike something you'd find squirming in a rain barrel and just about as easy to work with.

Well, since those days, striped bass, and subsequently its hybrids, have come a long way and so has my personal opinion of them.

Apparently the opinion of others has changed as well, as evidenced by the remarkable production and use of this species in recent years and the promise it poses for years to come. The evidence? Simply consider what you have heard today with respect to production gains and an ever expanding range.

Demand for striped bass and hybrid fingerlings continues to exceed the combined capabilities of the state and federal fish hatchery system. It seems that the species and its hybrids are going to be with us for some time to come.

Success of the striped bass program is directly associated with its use in reservoirs as a highly desirable sportfish and as a predator on shad and other large forage species. It also holds promise, as discussed earlier today, from a commercial standpoint.

The aquaculture industry in the United States stands poised for significant growth. Consider the fact that overall production has doubled in 5 years--300 million pounds in 1981. Catfish production alone increased to 64 million pounds the first 6 months of 1981, compared to a total production of 44 million pounds during 1977.

A New York based marketing research firm predicts that our aquaculture industry will grow from a $150 million industry in 1981 to a $1.6 billion industry by the end of the decade. Perhaps this is overly optimistic, but the point remains that, despite its growing pains, aquaculture is an emerging industry and the striper and its hybrids have a place in its emergence. Who knows, an individual may surface in the industry one day who will become to stripers or catfish what Frank Perdue is to chickens!

Whether cultured for sport or commercial purposes, more questions and needs seem to arise when working with a species than existed before you began. The striper is no exception. And although conferences of this type provide a means whereby professionals contribute to the state of the art, a number of questions concerning the striper remain unresolved.

To stimulate their being addressed at a future date and to provide additional enthusiasm for those studies already underway and identified here today, I list the following.

- broodstock development and strain evaluation
- pond fertilization techniques
- fry enumeration techniques
- factors affecting swim bladder inflation
- data on ammonia excretion rates of fingerlings
- additional work on intensive culture systems and on aeration systems
- continued study of the culture of striped bass in thermal effluents
- nutritional requirements of fry and fingerlings
- additional efficacious, cleared drugs and chemicals
- effects of intensity and periodicity of light
- vaccines
- effects of temperature and salinity on growth and survival
- determination of the levels of stress-induced hormones present in striped bass cultured in fresh vs. salt water
- gas supersaturation

To this point I have discussed the striper success story principally in terms of its contribution to inland sport fishery and management programs and touched upon its potential as a commercially viable species. But what about its status in its natural range and what problems, needs, and solutions have been identified thus far?

In recent years, there has been an acknowledged decline in striped bass populations on the mid and north Atlantic coasts. Not enough is known about standing stocks, sport and commercial harvest, natural mortality, spawning success, nor recruitment. More specifically, we need to better understand the impact of contaminants on spawning and early survival, upon activities in and around nursery grounds, and overfishing of the striper population by commercial and sport fishermen. Cooperative but under-funded state/federal research efforts are assisting in answering these questions—but is this enough?

Should we consider a major enhancement effort such as that undertaken for Pacific salmon? These efforts with stripers are underway elsewhere, why not the Chesapeake Bay for instance, an area one local writer described as "...a 10,000 year old puddle, left behind by the last ice age, plundered by everyone from Bluebeard to modern day pirates but always considered worthy of fighting over." Have we reached that point? Work by Consolidated Edison on the Hudson River suggests that the introduction of cultured striped bass into a estuarine wild population is feasible. Or perhaps states should follow California's lead and consider issuance of a striper stamp, with part of the resultant monies supporting enhancement efforts and the balance going for research.

It is obvious that many questions remain unanswered. But I'm confident that with time, the biologist, working in league with administrators and legislators, will ultimately arrive at a solution, amenable to all concerned, not the least of which is the species itself.

Paraphrasing a concluding earlier remark by Bob Stevens "What a fish story! What a fish!" I would add—what a future!

Thank you for the opportunity to summarize this program and personal thanks to Don Webster, Bill Sieling, Tony Mazzaccaro and others in Maryland who contributed so much to its development and organization.

Thanks as well to the speakers who gave of their time, knowledge, and effort in making this conference the success it has been. Last, but certainly not least, thanks to those of you who have supported this effort through your attendance today.

And Ken Allen, we missed your participation but are pleased that you're here today. We look forward to your working with us on next year's program.

Contributors

Ira Adelman
Department of Entomology, Fisheries and Wildlife, University of Minnesota. St. Paul, MN

Gary L. Burto
Tennessee Valley Authority Waste Heat Aquaculture Facility, Gallatin Steam Plant, Gallatin, TN

James M. Carlberg
Aquatic Systems Incorporated, La Jolla, CA

Charles M. Collins
Tennessee Valley Authority Wast Heat Aquaculture Facility, Gallatin Steam Plant, Gallatin, TN

Mike Freeze
Arkansas Aquatics, Inc., Lonoke, AR

J. Gabaudan
Department of Fisheries, Auburn University, Auburn, AL

Reginal M. Harrell
South Carolina Wildlife and Marine Resources Department, Bonneau, SC

James W. Kahrs
Osage Catfisheries, Inc., Osage Beach, MD

Jerome H. Kerby
North Carolina Cooperative Fishery Research Unit, Raleigh, NC

Michael J. Massingill,
Aquatic Systems Incorporated, La Jolla, CA

Joseph P. McCraren
U.S. Fish and Wildlife Service, Kearneysville, WV

Andrew J. Mitchell
U.S. Fish and Wildlife Service, Stuttgart, AR

Nick C. Parker
U.S. Fish and Wildlife Service, Marion AL

Ross L. Schweinforth
Tennessee Valley Authority, Waste Heat Aquaculture Facility, Gallatin Steam Plant, Gallatin, TN

Robert E. Stevens
U.S. Fish and Wildlife Service, Washington, D.C.

Dave Swartz
University of Maryland Sea Grant College, College Park, MD

C.J. Turner
Alabama Fish and Game Division, Montgomery, AL

Jon C. Van Olst
Aquatic Systems Incorporated, La Jolla, CA

John G. Woiwode
Department of Entomology, Fisheries and Wildlife, University of Minnesota, St. Paul, MN

L.C. Woods, III
Baltimore Gas and Electric Co., Baltimore, MD

Alex W. Wypyszinski
New Jersey Sea Grant Extension Service, New Brunswick, NJ

T. R. Zeigler
Zeigler Brothers, Inc., Gardners, PA